EXPERIMENTAL ORGANIC CHEMISTRY

CHEMISTRY 145

2014-2015
HOWARD UNIVERSITY EDITION

F. O. AYORINDE, O. BAKARE*, M. FELDMAN, J. FORTUNAK, A. M. HUDRLIK,
P. F. HUDRLIK, A. LEWIS, D. MOORE, R. C. ROSENBERG, J. W. WHEELER

*EDITOR

Editor: Joann Manos
Cover Designer: Doris Bruey

Experimental Organic Chemistry: 2014-2015 Howard University Edition

Permissions Department
Academx Publishing Services, Inc.
P.O. Box 208
Sagamore Beach, MA 02562
http://www.academx.com

Printed in the United States of America

ISBN-10: 1-60036-494-2
ISBN-13: 978-1-60036-494-5

TABLE OF CONTENTS

PREFACE

2013-2014 Edition

This laboratory manual is a result of our experience with several textbooks which we have used at Howard. We have found that commercial textbooks contain many experiments that we do not use and a lot of material that is not needed by our students. This manual contains background material and the directions for the experiments which will be performed in Chemistry 145 during the 2013-2014 academic year, as well as discussion of the techniques of the organic chemistry laboratory and instructions for writing notebooks and reports. This manual contains material contributed by the faculty who have taught the organic laboratory course, graduate teaching assistants, and students who have taken the course here at Howard. Based on our experience using previous versions of this manual during the past ten academic years, we have made many additions, revisions, and clarifications to the 2013-2014 edition.

We hope that you will learn and understand the techniques which are used routinely by professional chemists, and that you will enjoy the experience.

Chemistry 145 Faculty

F. O. Ayorinde
O. Bakare
M. Feldman
J. Fortunak
A. M. Hudrlik
P. F. Hudrlik
A. Lewis
D. Moore
R. C. Rosenberg
J. W. Wheeler

INTRODUCTION

OBJECTIVES

The object of the organic laboratory course is to become proficient in techniques used by practicing chemists. Organic laboratory differs from general chemistry laboratory in several ways. Numerical data compilation and mathematical manipulations are minimal in organic lab. The techniques which will be emphasized are

1. Separation and analysis of mixtures
2. Synthesis
3. Structure determination

Chemical and physical (instrumental) methods will be employed in a series of experiments listed on the schedule.

The student will be evaluated on work in the lab as well as on quizzes, exams and written laboratory reports. The student is expected to know what was done, why it was done, and what should be done in new situations. The final grade will be based on the sum of grades in several components, which will be listed by your instructor. We expect each student:

1. To know and follow correct laboratory safety and waste disposal procedures and practices at all times.

2. To know the names, structures, and relevant chemical and physical properties of all reactants, products, and solvents involved in each reaction carried out in the laboratory.

3. To be able to write a balanced equation for any chemical reaction, including redox reactions.

4. To understand and use the appropriate number of significant figures in measurements and calculations.

5. To understand and be able to calculate the percentage yield for both single and multi step reaction sequences. To understand and be able to identify the limiting reagent in a reaction or sequence of reactions.

6. To keep a satisfactory laboratory notebook. To know the difference between observations (data, measurements, results) and any conclusions based on these observations.

7. To know, understand, and be able to draw the mechanisms and stereochemical aspects of the reactions carried out in the laboratory insofar as they are discussed in this manual and/or in laboratory lecture.

8. To understand the practical aspects and theoretical basis of the various laboratory techniques (including spectroscopic techniques) and procedures used in the laboratory insofar as they are discussed in this manual and/or in laboratory lecture.

9. To be able to understand, interpret, and use chemical, physical, solubility, and flame test data, as well as IR, NMR, and mass spectra to determine the structure of an organic compound. Also, to be able to predict an organic compound's chemical properties, as well as the principal features in its IR, NMR, and mass spectra from its structure.

10. To be able to communicate your results in a concise well-written report.

SAFETY - General Considerations

The organic laboratory is a potentially dangerous place. However, the risks can be minimized if you are aware of the hazards and you follow safe laboratory practices. Your safety depends on your knowledge of possible dangers and your adherence to certain safety rules. The major hazards of the laboratory are the following:

Most organic compounds are <u>toxic</u> or <u>poisonous</u>. Many are carcinogenic, and some are corrosive. Dangerous exposure to chemicals can occur by inhalation or skin contact as well as by ingestion. Safe laboratory procedures and practices are designed to prevent you from contacting organic chemicals.

Most organic compounds are <u>flammable</u>, and flammable solvent vapors can travel a considerable distance along a laboratory bench or linger for some time in a sink. Safe laboratory procedures and practices are designed to minimize the chance of fire.

Some compounds are hazardous because of their <u>reactivity</u>; for example, reactivity with water (sodium metal, thionyl chloride) or oxygen (CS_2, phosphorus); or their instability (explosives).

<u>Apparatus</u> can be hazardous: Broken glass causes cuts; burners and hot glass cause burns; closed systems can explode.

MATERIAL SAFETY DATA SHEETS (MSDS)

Information about each chemical that is used in the Organic Chemistry Laboratory can be found in the Material Safety Data Sheet (MSDS) for that compound. These MSDS are housed on each floor of the Chemistry Building and are available for your inspection. They can also be found on the web (for example see, http://hq.msdsonline.com/howard3586). On the MSDS tab at this site type in the name of the chemical in the box on the "product" line and then choose "Starts With" as opposed to "Full Text" or "Contains a Match" or from the drop down menu).

Another convenient place to obtain MSDS's is at the website for Sigma-Aldrich, a chemical supplier: www.sigma-aldrich.com If your computer has not been to this site previously, you may be asked to indicate your country. As of this writing, the main page has a place to type in the "Product Name or No." for searching (e.g., type in the name of the compound of interest, and then click Go). If you don't see a place to type in the Product Name of No, then follow the links as if you planned to place an order, and at some point you should have an opportunity to search for the compound of interest. You will then be taken to a screen listing everything relevant to your search term (e.g., compounds with different isotopes) sold by Sigma-Aldrich. Click to expand the one of interest, and you will get a listing of all the available grades of that compound. Click on the relevant one (or one that appears to be fairly high purity such as Aldrich grade) to expand it and then click on the appropriate product number. The resulting screen will have information on that compound including a link to the MSDS. When you click on MSDS, you may be asked to log on. Ignore the log on and just click Next. The Sigma-Aldrich site only has the MSDS's for compounds they sell.

The MSDS for a compound contains very detailed information about the compound (structure, CAS (Chemical Abstracts) registry number, RTECS (Registry of Toxic Effects of Chemical Substances) number), its physical properties (such as melting point, boiling point, flash point), its reactivity (including incompatibility with other classes of chemicals, decomposition products), its toxicity (including health hazzards), and its safe handling (including storage, disposal, first aid following exposure, fire extinguishing methods, and spill clean up procedures).

As an example the MSDS for acetone is reproduced below. The "flash point" of a compound is the lowest temperature at which the compound in an open container will give off enough vapor so that a flame or spark brought near the surface will cause ignition. For acetone this temperature is −20 °C (see Section 3 and Section 9 of the MSDS), so acetone is extremely flammable and can certainly be ignited at room temperature by a nearby flame or spark. In terms of its toxicity, the LD50 of a compound is the dose, usually measured in mg of compound ingested per kg of body weight, that will cause 50% of the exposed individuals to die. The LD50 of a compound (see Section 11 of the MSDS) is usually available only for experimental animals such as mice, rats, or rabbits. The average human female weighs 120 lb or 55 kg while the average human male weighs 180 lb or 82 kg. Using the LD50 of 20 g/kg for skin exposure in rabbits, the calculated LD50 of acetone for skin exposure for the average human female would be

5280 g or 6.8 L, and for the average human male it would be 7940 g or 10.0 L. So if you accidently get a few drops of acetone on your skin, your exposure is well below the lethal level. However skin exposure to acetone may cause irritation, dryness, and inflammation. The potential health effects of exposure to acetone (even at well below lethal levels) are summarized in Section 3 of the MSDS. First aid measures for exposure to acetone are summarized in Section 4 of the MSDS.

MSDS for Acetone

Acetone, 99+%
ACROS95389

**** SECTION 1 - CHEMICAL PRODUCT AND COMPANY IDENTIFICATION****

MSDS Name: Acetone, 99+%
Catalog Numbers: AC177170000, AC177170010, AC177170025, AC177170200, AC177170250
Synonyms: 2-Propanone
Company Identification (Europe): Acros Organics N.V.
Janssen Pharmaceuticalaan 3a
2440 Geel, Belgium
Company Identification (USA): Acros Organics
One Reagent Lane
Fairlawn, NJ 07410
For information in North America, call: 800-ACROS-01
For information in Europe, call: 0032(0) 14575211
For emergencies in the US, call : CHEMTREC: 800-424-9300
For emergencies in Europe, call: 032(0) 14575299

**** SECTION 2 - COMPOSITION, INFORMATION ON INGREDIENTS ****

CAS#	Chemical Name	%	EINECS#
67-64-1	Acetone	99.0	200-662-2

Hazard Symbols: F
Risk Phrases: 11

**** SECTION 3 - HAZARDS IDENTIFICATION ****

EMERGENCY OVERVIEW

Appearance: clear, colorless. Flash Point: -20 deg C.
Danger! Extremely flammable liquid. May be harmful if absorbed through the skin. May cause reproductive effects based upon animal studies. May cause cardiac disturbances. Causes severe eye irritation. Causes severe respiratory tract irritation. May cause severe skin irritation. May cause digestive tract irritation with nausea, vomiting, and diarrhea.
Target Organs: Heart, central nervous system, reproductive system.

Potential Health Effects
Eye: Produces irritation, characterized by a burning sensation, redness, tearing, inflammation, and possible corneal injury.
Skin: Exposure may cause irritation characterized by redness, dryness, and inflammation.
Ingestion: May cause irritation of the digestive tract. May cause central nervous system depression, kidney damage, and liver damage. Symptoms may include: headache, excitement, fatigue, nausea, vomiting, stupor, and coma.
Inhalation: Inhalation of high concentrations may cause central nervous system effects characterized by headache, dizziness, unconsciousness and coma. Causes respiratory tract irritation. May cause motor incoordination and speech abnormalities. May cause narcotic effects in high concentration.
Chronic: Prolonged or repeated skin contact may cause dermatitis. Chronic inhalation may cause effects similar to those of acute inhalation.

**** SECTION 4 - FIRST AID MEASURES ****

Eyes: Immediately flush eyes with plenty of water for at least 15 minutes, occasionally lifting the upper and lower eyelids. Get medical aid immediately.
Skin: Immediately flush skin with plenty of soap and water for at least 15 minutes while removing contaminated clothing and shoes. Get medical aid if irritation develops or persists. Wash clothing before reuse.
Ingestion: Do NOT induce vomiting. If victim is conscious and alert, give 2-4 cupfuls of milk or water. Never give anything by mouth to an unconscious person. Get medical aid immediately.
Inhalation: Get medical aid immediately. Remove from exposure to fresh air immediately. If not breathing, give artificial respiration. If breathing is difficult, give oxygen.
Notes to Physician: Treat symptomatically and

**** SECTION 5 - FIRE FIGHTING MEASURES ****

General Information:
Containers can build up pressure if exposed to heat and/or fire. As in any fire, wear a self-contained breathing apparatus in pressure-demand, MSHA/NIOSH (approved or equivalent), and full protective gear. Vapors can travel to a source of ignition and flash back. During a fire, irritating and highly toxic gases may be generated by thermal decomposition or combustion. Will burn if involved in a fire. Use water spray to keep fire-exposed

containers cool. Vapor may cause flash fire. Water may be ineffective. Material is lighter than water and a fire may be spread by the use of water. Vapors may be heavier than air. They can spread along the ground and collect in low or confined areas.

Extinguishing Media:

For small fires, use dry chemical, carbon dioxide, water spray or alcohol-resistant foam. Water may be ineffective. For large fires, use water spray, fog or alcohol-resistant foam. Do NOT use straight streams of water. Cool containers with flooding quantities of water until well after fire is out.

**** SECTION 6 - ACCIDENTAL RELEASE MEASURES ****

General Information:

Use proper personal protective equipment as indicated in Section 8.

Spills/Leaks:

Absorb spill with inert material, (e.g., dry sand or earth), then place into a chemical waste container. Avoid runoff into storm sewers and ditches which lead to waterways. Clean up spills immediately, observing precautions in the Protective Equipment section. Scoop up with a nonsparking tool, then place into a suitable container for disposal. Remove all sources of ignition. Provide ventilation.

**** SECTION 7 - HANDLING and STORAGE ****

Handling:

Wash thoroughly after handling. Wash hands before eating. Use only in a well ventilated area. Use spark-proof tools and explosion proof equipment. Do not breathe dust, vapor, mist, or gas. Do not get in eyes, on skin, or on clothing. Empty containers retain product residue, (liquid and/or vapor), and can be dangerous. Avoid contact with heat, sparks and flame. Do not ingest or inhale. Do not pressurize, cut, weld, braze, solder, drill, grind, or expose empty containers to heat, sparks or open flames.

Storage:

Keep away from heat, sparks, and flame. Keep away from sources of ignition. Do not store in direct sunlight. Store in a tightly closed container. Store in a cool, dry, well-ventilated area away from incompatible substances. Flammables-area.

**** SECTION 8 - EXPOSURE CONTROLS, PERSONAL PROTECTION ****

Engineering Controls:

Use process enclosure, local exhaust ventilation, or other engineering controls to control airborne levels below recommended exposure limits. Facilities storing or utilizing this material should be equipped with an eyewash facility and a safety shower.

Exposure Limits

Chemical Name	ACGIH	NIOSH	OSHA - Final PELs
Acetone	500 ppm; 750 ppm STEL	250 ppm TWA; 590 mg/m^3 TWA 2500 ppm IDLH (lower explosive level)	1000 ppm TWA; 2400 mg/m^3 TWA

OSHA Vacated PELs:

Acetone: 750 ppm TWA; 1800 mg/m3 TWA

Personal Protective Equipment

Eyes: Wear appropriate protective eyeglasses or chemical safety goggles as described by OSHA's eye and face protection regulations in 29 CFR 1910.133 or European Standard EN166.

Skin: Wear appropriate gloves to prevent skin exposure.

Clothing: Wear appropriate protective clothing to prevent skin exposure.

Respirators:

Follow the OSHA respirator regulations found in 29CFR 1910.134 or European Standard EN 149. Always use a NIOSH or European Standard EN 149 approved respirator when necessary.

**** SECTION 9 - PHYSICAL AND CHEMICAL PROPERTIES ****

Physical State: Liquid
Appearance: clear, colorless
Odor: Not available.
pH: Not available.
Vapor Pressure: 247 mbar @ 20 C
Vapor Density: 2.0
Evaporation Rate: 7.7 (n-Butyl acetate=1)
Viscosity: 0.32 mPas 20 de
Boiling Point: 56 deg C @ 760.00mm Hg
Freezing/Melting Point: -94 deg C
Autoignition Temperature: 465 deg C (869.00 deg F)
Flash Point: -20 deg C (-4.00 deg F)
NFPA Rating: (est.) Health: 1; Flammability: 3; Reactivity: 0
Explosion Limits, Lower: 2.50 vol %
Upper: 13.00 vol %
Decomposition Temperature:
Solubility: soluble
Specific Gravity/Density: .7910g/cm^3
Molecular Formula: C_3H_6O
Molecular Weight: 58.08

**** SECTION 10 - STABILITY AND REACTIVITY ****

Chemical Stability: Stable.

Conditions to Avoid:

Mechanical shock, incompatible materials, ignition sources, moisture, excess heat, exposure to moist air or water.

Incompatibilities with Other Materials:

Bases, oxidizing agents, reducing agents, moisture. Forms explosive mixtures with hydrogen peroxide, acetic acid, nitric acid, nitric acid+sulfuric acid, chromic anhydride, chromyl chloride, nitrosyl chloride, hexachloromelamine, nitrosyl perchlorate, perchlorate, permonosulfuric acid, thiodiglycol+hydrogen peroxide, potassium ter-butoxide, sulfur dichloride, 1-methyl-1,3-butadiene, bromoform, carbon, air, chloroform, thitriazylperchlorate.

Hazardous Decomposition Products: Carbon monoxide, carbon dioxide.

Hazardous Polymerization: Will not occur.

**** SECTION 11 - TOXICOLOGICAL INFORMATION ****

RTECS#:

CAS# 67-64-1: AL3150000

LD50/LC50:

CAS# 67-64-1: Inhalation, rat: LC50 =50100 mg/m3/8H; Oral, mouse: LD50 = 3 gm/kg; Oral, rabbit: LD50 = 5340 mg/kg; Oral, rat: LD50 =5800 mg/kg; Skin, rabbit: LD50 = 20 gm/kg.

Carcinogenicity: Acetone -ACGIH: A4 - Not Classifiable as a Human Carcinogen

Epidemiology: No information available.

Teratogenicity:

No developmental toxic effects were seen in rats and mice exposed to atmospheric concentrations of acetone vapour up to 11,000 ppm and 6000 ppm respectively 6 hr/day for 7 day.

Reproductive Effects:

Fertility: post-implantation mortality. Ihl, mam:TCLo=31500 ug/m^3/24H (1-13D preg)

Neurotoxicity: No information available.

Mutagenicity:

Cytogenetic analysis: hamster fibroblast, 40 g/L Sex chromosome loss/non-disjunction: S.cerevisiae, 47600 ppm Salmonella typhimurium TA92, TA94, TA98, TA100, TA1535, TA1537 with metabolic activation negative. Chinese hamster fibroblast (24 hr) without metabolic activation induced chromosomal aberrations.

Other Studies: None.

**** SECTION 12 - ECOLOGICAL INFORMATION ****

Ecotoxicity:

Fish: Rainbow trout: LC50 = 5540 mg/L; 96 Hr.; Static conditions; 13 degrees CFish: Fathead Minnow: LC50 = 7280-8120 mg/L; 96 Hr.; Flow-through conditions; 24-26 degrees CRainbow trout LC50:5540 mg/L/96H Sunfish (tap water), death at 14250 ppm/24H Mosquito fish (turbid water) TLm=13000 ppm/48H Harlequin fish LC50:5700 ppm/24H Daphnia magna ED50:10 mg/l/24,48H Brine shrimp LD50:2100 mg/l/24,48H at 24C

**** SECTION 13 - DISPOSAL CONSIDERATIONS ****

Chemical waste generators must determine whether a discarded chemical is classifas a hazardous waste.
US EPA guidelines for the classification determination are listed in 40 CFR Part
Additionally, waste generators must consult state and local hazardous waste regu ensure complete and accurate classification.
RCRA P-Series: None listed.
RCRA U-Series: CAS# 67-64-1: waste number U002;
(Ignitable waste).

**** SECTION 14 - TRANSPORT INFORMATION ****

US DOT

Shipping Name:	ACETONE
Hazard Class:	3
UN Number:	UN1090
Packing Group:	II

Canadian TDG

Shipping Name:	ACETONE
Hazard Class:	3
UN Number:	UN1090

Other Information: FLASHPOINT -20 C

**** SECTION 15 - REGULATORY INFORMATION ****

US FEDERAL

TSCA

CAS# 67-64-1 is listed on the TSCA inventory.

Health & Safety Reporting List

None of the chemicals are on the Health & Safety Reporting List.

Chemical Test Rules

None of the chemicals in this product are under a Chemical Test Rule.

Section 12b

None of the chemicals are listed under TSCA Section 12b.

TSCA Significant New Use Rule

None of the chemicals in this material have a SNUR under TSCA.

SARA

Section 302 (RQ)

CAS# 67-64-1: final RQ = 5000 pounds (2270 kg)

Section 302 (TPQ)

None of the chemicals in this product have a TPQ.

SARA Codes

CAS # 67-64-1: acute, chronic, flammable, sudden release of pressure.

Section 313

No chemicals are reportable under Section 313.

Clean Air Act:
This material does not contain any hazardous air pollutants.
This material does not contain any Class 1 Ozone depletors.
This material does not contain any Class 2 Ozone depletors.

Clean Water Act:
None of the chemicals in this product are listed as Hazardous Substances under the CWA.
None of the chemicals in this product are listed as Priority Pollutants under the CWA.
None of the chemicals in this product are listed as Toxic Pollutants under the CWA.

OSHA:
None of the chemicals in this product are considered highly hazardous by OSHA.

STATE

Acetone can be found on the following state right to know lists: California, New Jersey, Florida, Pennsylvania, Minnesota, Massachusetts.

California No Significant Risk Level:
None of the chemicals in this product are listed.

European/International Regulations

European Labeling in Accordance with EC Directives
Hazard Symbols: F
Risk Phrases: R 11 Highly flammable.
Safety Phrases:
S 9 Keep container in a well-ventilated place.
S 16 Keep away from sources of ignition - No smoking.
S 23 Do not inhale gas/fumes/vapour/spray.
S 33 Take precautionary measures against static discharges.

WGK (Water Danger/Protection)
CAS# 67-64-1: 0

United Kingdom Occupational Exposure Limits
CAS# 67-64-1: OES-United Kingdom, TWA 750 ppm TWA; 1810 mg/m^3 TWA
CAS# 67-64-1: OES-United Kingdom, STEL 1500 ppm STEL; 3620 mg/m^3 STEL

Canada
CAS# 67-64-1 is listed on Canada's DSL/NDSL List.
This product has a WHMIS classification of B2, D2B.
CAS# 67-64-1 is not listed on Canada's Ingredient Disclosure List.

Exposure Limits
CAS# 67-64-1: OEL-AUSTRALIA:TWA 500 ppm (1185 mg/m3);STEL 1000 ppm
OEL-AUSTRIA:TWA 750 ppm (1780 mg/m3)
OEL-BELGIUM:TWA 750 ppm (1780 mg/m3);STEL 1000 pp
OEL-CZECHOSLOVAKIA:TWA 800 mg/m3;STEL 4000 mg/m^3
OEL-DENMARK:TWA 250 ppm (600 mg/m3)
OEL-FINLAND:TWA 500 ppm (1200 mg/m3);STEL 625 ppm (1500 mg/m3)
OEL-FRANCE:TWA 750 ppm (1800 mg/m3)
OEL-GERMANY:TWA 1000 ppm (2400 mg/m3)
OEL-HUNGARY:TWA 600 mg/m3;STEL 1200 mg/m3
OEL-INDIA:TWA 750 ppm (1780 mg/m3);STEL 1000 ppm (2375 mg/m3)
OEL-JAPAN:TWA 200 ppm (470 mg/m3)
OEL-THE NETHERLANDS:TWA 750 ppm (1780 mg/m3) JAN9
OEL-THE PHILIPPINES:TWA 1000 ppm (2400 mg/m3)
OEL-POLAND:TWA 200 mg/m3
OEL-RUSSIA:TWA 200 ppm;STEL 200 mg/m3
OEL-SWEDEN:TWA 250 ppm (600 mg/m3);STEL 500 ppm (1200 mg/m3)
OEL-SWITZERLAND:TWA 750 ppm (1780 mg/m3)
OEL-TURKEY:TWA 1000 ppm (2400 mg/m3)
OEL-UNITED KINGDOM:TWA 750 ppm (1810 mg/m3);STEL 1250 ppm
OEL IN BULGARIA, COLOMBIA, JORDAN, KOREA check ACGIH TLV
OEL IN NEW ZEALAND, SINGAPORE, VIETNAM check ACGI TLV

**** SECTION 16 - ADDITIONAL INFORMATION ****

MSDS Creation Date: 7/27/1999 Revision #4 Date: 8/02/2000

The information above is believed to be accurate and represents the best information currently available to us. However, we make no warranty of merchantability or any other warranty, express or implied, with respect to such information, and we assume no liability resulting from its use. Users should make their own investigations to determine the suitability of the information for their particular purposes. In no way shall the company be liable for any claims, losses, or damages of any third party or for lost profits or any special, indirect, incidental, consequential or exemplary damages, howsoever arising, even if the company has been advised of the possibility of such damages.

SAFE LABORATORY PRACTICES

EYE PROTECTION

One of the most important safety precautions for any chemistry laboratory is protecting your eyes. Therefore, when work is going on (including check in and check out), everyone will wear either safety goggles or safety glasses (whether you are working or not - your neighbor may be). A 2-point penalty will be assessed every time you are caught without eye protection. Repeat offenders will be removed from the laboratory. If you have finished your experimental work and wish to complete your report without your goggles, you may do so in the hall outside the lab or in a vacant classroom.

Contact lens wearers: According to the Committee on Chemical safety of the American chemical Society (http://www.pubs.acs.org/cen/safety/19980601.html) and the American Optometric Association (http://www.aoanet.org/ia-cl-ind-environ.html) there is no evidence of additional risk to persons wearing contact lenses in the laboratory as long as they use approved eye protective devices, which in the case of the Organic Chemistry laboratory are safety goggles or safety glasses.

SKIN PROTECTION & APPROPRIATE CLOTHING

Preventing chemicals from contacting your skin is also one of the most important aspects of safety in the Organic Chemistry laboratory. Exposed skin is to be avoided. Your clothing must completely cover your torso, legs, and feet. Open toe shoes, sandals, shorts, or short skirts are **not** permitted. You may wear a laboratory coat. However, laboratory coats do not completely cover your legs. Loose, long hair and very loose clothing or loose jewelry is also **not** permitted. You may be removed from the Organic Chemistry laboratory if you are inappropriately dressed.

EMERGENCY PROCEDURES

(From: "Safety In Academic Chemistry Laboratories," Committee on Chemical Safety, American Chemical Society, Sixth Edition, pp 32-36, used by permission)

Chemistry laboratories have a variety of equipment to be used in case an accident occurs. Each person in a laboratory must be familiar with the locations of, and procedures for using this equipment. The safety equipment includes an eyewash fountain, safety shower, fire extinguisher, and an alarm or telephone system. The laboratory has a plan for everyone to follow if an evacuation is ever necessary. Be sure that you know the main and alternate evacuation routes as well as the procedures for accounting for each person in the laboratory.

1. General Information

As much as possible, follow procedures that have been established and practiced. When helping another person, remember to evaluate the potential danger to yourself before taking action. When an emergency occurs, the following actions are recommended:

(A) Report the nature and location of the emergency to the instructor (or laboratory technician) and, if necessary (they are incapacitated) to the appropriate fire (Howard University Security Emergency: 202-806-1100 (x61100), District of Columbia Emergency: 9-911) or medical (Howard University Student Health Center: 202-806-7540 (x67540), District of Columbia Emergency: 9-911), and give your name, telephone number, and address. Tell where you will meet the emergency vehicle. If individuals are involved, report how many; whether they are unconscious, buried, or trapped; whether an explosion has occurred; and whether there is or has been a chemical or electrical fire.

(B) Tell others in the area about the nature of the emergency

(C) Do not move any injured person unless they are in immediate danger from chemical exposure or fire. Keep them warm. Unnecessary movement can severely complicate neck injuries and fractures.

(D) Meet the ambulance or fire crews at the place you indicated. Send someone else if you cannot go.

(E) Do not make any other telephone calls unless they directly relate to the control of the emergency.

2. Fires

2.1 Fire Prevention

The best way to fight a fire is to prevent it. Fires can be prevented and their severity considerably reduced by proper housekeeping and by thoughtful reflection about what you are doing. This includes prompt removal of waste, separation of flammable liquids from combustible material such as cardboard boxes and paper towels, storage only of limited quantities of flammable material, and unobstructed aisles and exits. Stand back, take a look and ask: "Are there any frayed wires? Is a sparking motor stirrer being used to stir a flammable liquid? Are those bottles too close to the edge of the bench? Is the work space cluttered? Do I understand each of the potential hazards in what I am about to do? Am I prepared in advance to take preventative steps?"

2.2 Dealing with a Fire

When a fire occurs the following actions are recommended:

(A) A fire contained in a small vessel can usually be suffocated by covering the vessel. Do not pick up the vessel. Do not cover with dry towels or cloths. Do, for example, use a watch glass to suffocate a fire in a beaker. Remove nearby flammable materials to avoid spread of the fire. Do NOT attempt to use a fire extinguisher unless you have been trained in its use and know that it is likely to extinguish the fire.

(B) If the fire is burning over an area too large for the fire to be suffocated quickly and simply, all persons should evacuate the area except those trained and equipped to fight fires. Do not use elevators to leave the building; use the stairs.

(C) Activate the fire alarm. Notify co-workers and the instructor (or laboratory technician). If they are incapacitated call the fire department (District of Columbia Emergency: 9-911, and Howard University Security Emergency: 202-806-1100 (x61100)). As much as possible, follow evacuation procedures that have been established and that you have practiced during prior fire drills.

(D) If you have been trained in the use of a fire extinguisher, fight the fire from a position from which you can escape, and only if you are confident that you will be successful. Small fires just starting often can be extinguished, but not always. If not extinguished, a fire can quickly threaten your life and that of your co-workers. Remember it is easy to underestimate a fire.

(E) Toxic gases and smoke may be present during a fire., and those persons trying to contain the fire must avoid breathing gases and smoke. These fires should be fought only by properly trained personnel equipped with self-contained breathing apparatus.

(F) Smother fires involving very reactive metals with powdered graphite or with a fire extinguisher designed for metal fires. Carbon dioxide and the usual dry chemical fire extinguishers will intensify fires of alkali, alkaline earth and certain other metals, including aluminum, magnesium, zirconium, hafnium, thorium, and uranium.

(G) Fire fighters should be informed what chemicals are involved or which chemicals may become involved. A current inventory list is required and a copy should be readily available outside the work area. Ideally, but only if the information is kept up to date, laboratory doors should be posted with the National Fire Protection Association diamond, which provides information to the fire fighters.

(H) Fire involving chemicals increases the possibility of explosions. Special care should be taken in advance of a fire to keep the potential for fire or excessive heat as far as possible from volatile solvents, compressed gas cylinders, reactive metals, and explosive compounds.

(I) Immediately after the fire, all extinguishers that were used should be recharged or replaced with full ones.

2.3 Personal Injuries Involving Fires

If a person's clothing is on fire, that person should use the safety shower. If the shower is not readily available, douse the individual with water or wrap the person in a coat, blanket, or whatever is available to extinguish the fire and roll the person on the floor. Fire blankets should be used with caution because wrapping the body can force flames toward the face and neck. Quickly remove any clothing contaminated with chemicals. To prevent contamination of the eyes use caution when removing pullover shirts or sweaters. Douse with water to remove heat and place clean, wet, cold cloths on burned areas. Wrap the injured person to avoid shock and exposure. Get medical attention promptly. If the instructor or laboratory technician are incapacitated call Howard University Student Health Center: 202-806-7540 (x67540), or Howard University Security Emergency: 202-806-1100 (x61100), or District of Columbia Emergency: 9-911.

3. Chemical Spills on Personnel

For spills covering small amounts of skin, immediately flush with flowing water for no less than 15 minutes. If there is no visible burn, wash with warm water and soap, removing any jewelry to facilitate removal of any residual materials. Check the MSDS to see if any delayed effects should be expected. It is advisable to seek medical attention for even minor chemical burns (Go to the Howard University Student Health Center: 202-806-7540 (x67540), Medical Arts Building, 2139 Georgia Ave NW (Georgia Ave. and W Street NW, Suite 201 - second floor).

For spills on clothes, don't attempt to wipe or flush off the clothes. Quickly remove all contaminated clothing, shoes, and jewelry while using the safety shower. Seconds count, and no time should be wasted because of modesty. Be careful not to spread the chemical on the skin, or especially in the eyes. To prevent contamination of the eyes use caution when removing pullover shirts or sweaters; it may be better to cut the garments off. Immediately flood the affected body area with tempered water for at least 15 minutes. Resume if pain returns. Do not use creams, lotions, or salves. Get medical attention as soon as possible (go to the Howard University Student Health Center: 202-806-7540 (x67540), , Medical Arts Building, 2139 Georgia Ave NW (Georgia Ave. and W Street NW, Suite 201 - second floor)). Launder contaminated clothes separately from other clothing or discard, as recommended in the MSDS. Never work with chemicals in a laboratory that is not equipped with a safety shower.

Your instructor should make certain that the medical personnel understand exactly what chemicals are involved and that medical personnel (including physicians, nurses, and paramedics) recognize and use proper treatment for that exposure. Preferably this should be done in advance of any potential emergency. The exact chemical name should be supplied. For example, exposure to hydrochloric acid is very different medically than exposure to hydrofluoric acid, yet both are sometimes called acids.

For splashes into the eye, immediately flush the eye with tempered potable water from a gently flowing source for at least 15 minutes. Hold the eyelids away from the eyeball, move the eye up and down and sideways to wash thoroughly behind the eyelids. An eyewash fountain should be used, but if one is not available, injured persons should be placed on their backs and water gently poured into the corners of their eyes for at least 15 minutes. Never work with chemicals in a laboratory that is not equipped with an eyewash fountain. First aid for the eyes must always be followed by prompt treatment by a member of a medical staff or an ophthalmologist acquainted with the management of chemical injuries to the eyes (go to the Howard University Student Health Center: 202-806-7540 (x67540), Medical Arts Building, 2139 Georgia Ave NW (Georgia Ave. and W Street NW, Suite 201 - second floor)).

4. Other Accidents Involving Personal Injury

Anyone overcome with smoke or fumes should be removed to uncontaminated air and treated for shock. Remember to evaluate and describe the possibility of harm to the rescuer before the rescuer enters or continues to remain in a toxic environment. If hazardous chemicals are ingested, follow the first aid treatment shown on the label or in the MSDS. Never give anything by mouth to an unconscious person. Attempt to learn exactly what substances were ingested and inform the medical staff (while the victim is en route to a hospital, if possible). Refer to the MSDS for information regarding treatment.

If the injured person is not breathing, provide mouth-to-mouth resuscitation. If there is no pulse, administer cardiopulmonary resuscitation (CPR).

If an individual is bleeding severely, control the bleeding by compressing the wound with a cloth or whatever is available. Elevate the injury above the level of the heart. If blood is spurting, place a pad directly on the cut and apply firm pressure. Take reasonable precautions to avoid contact with blood. Wrap the injured person to avoid shock and get immediate medical attention. Notify the instructor or laboratory technician or if they are incapacitated call the Howard University Student Health Center: 202-806-7540 (x67540) or Howard University Security Emergency: 202-806-1100 (x61100), District of Columbia Emergency: 9-911). A pressure pad should be applied firmly on the wound. Tourniquets should only be used by persons trained in first aid.

Do not touch a person in contact with a live electrical circuit. Disconnect the power first or the rescuer may be seriously injured.

5. Chemical Spills

All spills should be cleaned up promptly, efficiently, and properly. Notify the person in charge (instructor or laboratory technician) for help. If they are incapacitated call Howard University Environmental Health and Safety: 202-806-1006 (x61006), 202-806-1007 (x61007) or Howard University Security Emergency: 202-806-1100 (x61100), or District of Columbia Fire Department Hazardous Materials Unit: 9-911). Warn all individuals at risk of involvement. Not only will they be spared exposure to the hazard, but also the spread of the hazard will be minimized. For chemical spills on the skin or in the eyes, treatment must begin immediately. Often the volume spilled is not as important as the toxicity of the substance.

Containing the spill rapidly is a primary concern, because the smaller the area involved, the easier the cleanup. Diking of liquids by surrounding the involved area with an absorbent retaining material is recommended. After the spill has been contained, it can be cleaned up with appropriate materials. Commercially available or homemade spill control kits can be useful here.

If there is no fire hazard and the material is not believed to be volatile or toxic, clean it as directed by the instructor or laboratory technician. Refer to the appropriate MSDS. To facilitate cleaning up liquids, use an absorbent material that will neutralize the liquids if possible (trisodium phosphate, sand followed by sodium bicarbonate solution or powder for acids, sodium thiofulfate solution for bromine, etc.). Commercial absorbents (e.g., Oil-Dri and Zorb-All), vermiculite, or small particles (about 30 mesh) of Kitty Litter or other satisfactory clay absorbents can be used. Dry sand is less effective. A dustpan and brush should be used, and protective gloves should be worn. While wearing gloves, clean the contaminated area with soap and water and mop it dry. If the spill is on the floor, some absorbent should be sprinkled on the spot to prevent slipping. Note that vermiculite and some other absorbents create a slipping hazard when wet.

If a volatile, flammable, or toxic material is spilled immediately warn everyone to extinguish flames and turn of spark producing equipment such as brush type motors. Shut down all equipment and leave the area until it is decontaminated. Clothing contaminated by spills or splashes should be immediately removed to prevent skin penetration. The instructor (or laboratory technician) will be responsible for designating the extent of evacuation and the proper cleanup procedure.

Many small liquid spills (<100 mL) can be absorbed with paper towels, sand, or an absorbent. However, paper towels can increase the surface area and rate of evaporation, increasing the fire hazard. Most spills of solids can be brushed up and disposed of in appropriate solid waste containers, but care must be exercised to avoid reactive combinations. Do not leave paper towels or other materials used to clean up a spill in open trash cans in the work area.

After cleanup, all materials, including paper towels used in the cleanup, must be disposed of as wastes. Be particularly careful that flammable liquids absorbed during cleanup do not present a continuing fire hazard.

OTHER GOOD LABORATORY PRACTICES

Clean up spills immediately. Keep paper towels handy for this purpose. Use your cloth towel for cleaner tasks. Keep your melting point tube in a large beaker to minimize the possibility of an oil spill.

Help prevent flooding. Turn off the water when you are finished with it. Don't throw junk in the sink or drain trough. Don't leave glassware or brushes in the sink. Wet floors are slippery - walk with care, do not run.

Try to avoid taking more reagent than you need, but if you do, don't pour it back into the reagent bottle. See if someone else can use it, or dispose of it in the appropriate waste container.

Prevent traffic jams - avoid bottlenecks between hoods and sinks.

Don't pile books and coats on the air conditioning units. If it should be necessary to adjust the temperature, turn the thermostat knob slightly. [Leave the switch on AUTO.] If you open a window, be sure to close it later.

The water faucets can be used as a source of water (for a condenser) or as an aspirator (for a suction filtration). Do not try to use one faucet for both purposes at once. Although condensers need only a small water flow, aspirators work best if they are turned on all the way. Splashing can usually be minimized by attaching a 8 - 10-inch piece of stiff tubing to the outlet. If many aspirators are used at once, keep an eye on the sink - it may fill up and overflow. Break the vacuum before turning off the aspirator.

Keep your drawer closed. Open drawers can be a traffic obstacle and may get filled with water if your neighbor has an accident with his condenser tubing. Lock your drawer if you go away for any period of time. [But don't lock up common equipment like burners, heating mantles, clamps, or rubber tubing.]

Scientific instruments are expensive and can be difficult to repair. Keep them clean and dry, and wipe up spills promptly. If you think any of the instruments are not functioning properly, bring this to the attention of your instructor immediately.

WASTE DISPOSAL

A waste is a material (solid or liquid) that will no longer be used and is to be discarded. Special methods or procedures must be followed to dispose of wastes produced in the organic chemistry laboratory. The regulations of the United States Environmental Protection Agency (EPA) and the District of Columbia severely limit the wastes that can be disposed of into the sanitary sewer system (ie. down the drain). Special containers are available in the organic laboratory for disposal of each type of waste generated in this course. They are:

Organic Waste: A blue plastic 10 gallon jug is located in one of the hoods. Dispose of all liquid organic wastes such as organic solvents, liquid organic solutions (including acetone used to dry glassware), and liquid organic compounds into this container.

Broken Glassware: A cardboard receptacle lined with a thick plastic bag is located near the door. Dispose of all broken glassware, used pasteur pipets, used melting point capillaries, and any other waste glass into this container. Do not put paper towels or other solid wastes into this container.

Solid Waste: A large trash container with a lid is located near the door. Dispose all other solid waste into this container. Examples include used paper and paper towels, used weighing boats, used filter paper, used cloth towels, used sodium sulfate, used boiling chips, used glass wool, and used calcium chloride ($CaCl_2$). Waste solid organic compounds should be dissolved in acetone and this solution should be poured into the Organic Waste container.

Neutralized aqueous solutions of inorganic acids and bases may be poured down the drain, followed by lots of water. Examples include neutralized aqueous solutions of hydrochloric acid (HCl), nitric acid (HNO_3), sulfuric acid (H_2SO_4), phosphoric acid (H_3PO_4), sodium hydroxide (NaOH), potassium hydroxide (KOH), sodium carbonate (Na_2CO_3), sodium bicarbonate ($NaHCO_3$), aqueous solutions of both sodium chloride (NaCl, including saturated sodium chloride solutions) and sodium sulfate (Na_2SO_4).

If you are not sure how to properly dispose of a particular waste, ask your instructor or the laboratory technician.

SAFETY GUIDELINES

The following safety rules must be followed at all times:

1. No one may work in the lab unless an instructor is present.

2. In case of emergency (fire, injury, corrosive spill, etc.) notify the instructor or other responsible person immediately.

3. Safety goggles must be worn at all times in the laboratory. No bare feet, open shoes, or sandals. No loose, long hair, very loose clothing, or very loose jewelry.

4. No eating, drinking, or smoking in the laboratory.

5. No flames are permitted unless approved by the lab instructor.

6. No solids of any kind, including matches, should be put in the sink; discarded labels, paper, glass wool, etc. should be put in the solid waste container.

7. Be neat. Wipe up spilled chemicals immediately. Keep balances and instruments clean.

8. Avoid contact with chemicals on your skin. Wash immediately if chemicals come in contact with your skin or clothing.

9. Know the location of fire extinguishers, safety showers, eyewash fountains, fire blankets, and first aid kit, and know how to use them.

10. Before doing any laboratory work, read the safety information on the inside front cover and in this manual.

11. Before starting any new experiment make sure you understand the potential hazards, any necessary safety precautions, and proper waste disposal procedures.

SAFETY GUIDELINES FORM

Complete this form and return it to your laboratory instructor.

Name (print legibly)________________________________

I.D. ________________________________

Section ________________________________

Desk No. ________________________________

Safety Rules and Guidelines

I understand that the organic chemistry laboratory is a potentially dangerous environment, and that my safety and the safety of others depend on my knowledge of possible dangers and of safe laboratory practices. By my signature below, I acknowledge the following:

- That I have read the safety information in this manual.
- That I will follow the safety guidelines and procedures in this manual at all times.
- That I understand that if I do not comply with the safety guidelines and procedures: 1) I may be immediately removed from the laboratory; 2) I may not receive credit for the experiment; 3) I may not be permitted to make up the experiment; or 4) I may receive a failing grade.
- That I am expected to comply with all laboratory safety guidelines, including additional safety instructions given by my instructor/professor prior to or during laboratory experiments that I may perform.
- That failure to comply with <u>all</u> laboratory safety guidelines and procedures may also be violations of academic and conduct policies and may subject me to disciplinary procedures outlined in the Howard University H- Book and the Student Code of Conduct.

Signature ________________________________ Date ________________

ORGANIC LABORATORY CHECK-IN PROCEDURE

1. Each laboratory bench location has a card with a desk number and list of equipment on it.

2. Print your name and ID number on the card.

3. Empty your drawer. Place everything from the drawer on the bench top. Clean the drawer: Wipe it with a paper towel, and if necessary remove the drawer and dump the trash in the solid waste container. Refer to the drawings (after the equipment list) and arrange the glassware and equipment in the order listed on the next page. Your instructor or teaching assistant will help you identify equipment and glassware.

4. Draw a line through or otherwise mark every item that you have.
   You do not need a triangular file or a wing tip.
   The tincture bottles are not important.
   If one of the glass funnels (fluted or Bunsen) has a short stem, that's OK.
   The Büchner funnel (porcelain) should have a rubber stopper attached.
   Each separatory funnel should have a matching stopper (glass or teflon).
   If the drying tubes have calcium chloride (white granular material) in them, clean them out (remove dry solid, then wash with water).
   The melting-point tube will probably have oil in it; if so, keep it upright in the large beaker.

5. Wash any dirty glassware using detergent and water. Rinse well with water. After draining any water, rinse the glassware with a small amount of acetone, and then allow the glassware to air dry. Collect the acetone in a beaker and then dispose the acetone into the waste organic container in the hood.

   Neatly place all glassware and equipment back in your drawer along with the card listing the equipment.

6. Lock your drawer with your lock or with the lock provided if you have not yet obtained your own lock.

7. After reading the SAFETY sections of this manual, complete and sign the "Safety Rules" Form and return the signed form to your instructor.

EQUIPMENT FOR ORGANIC CHEMISTRY LAB, 145

(**BOLD** numbers refer to illustration of glassware)

1 beakers, 100 mL (2)
1 beakers, 250 mL (2)
1 beaker, 1000 mL (1)
2 Erlenmeyer flasks, 125 mL (2)
2 Erlenmeyer flasks, 250 mL (2)
3 filter flask, 250 mL (1)
4 graduated cylinder, 25 mL (1)
5 graduated cylinder, 100 mL (1)
6 glass funnels (Bunsen (1) and fluted (1), one may have short stem)
7 Buchner funnel (porcelain, with rubber stopper) (1)
8 separatory funnel, 250 ml (including stopper) (1)

20 wire gauze (1)
21 glass stirring rod (1)
spatula (1)
thermometer, 250°C (1)
test tube rack (1)
test tubes, 3 in. (2)
test tubes, 6 in. (6)
test tubes, 8 in. (1)
goggles (1 pr)
cloth towel (1)
misc. corks, pipettes, vials

<u>19/22 standard taper glassware (1 ea)</u>

9 round bottom flask, 25 mL
9 round bottom flask, 50 mL
9 round bottom flask, 100 mL
10 vacuum adapter
11 connecting adapter (Y-shaped, 2 male & 1 female joints)
12 outlet adapter (small - with neoprene fitment)
12 neoprene fitment (usually attached to outlet adapter)
13 Claisen adapter (1 male & 2 female joints)
14 West condenser (thin)
15 distillation column (fat)
16 separatory funnel, 125 mL
17 stopper, 19/22
18 round bottom flask, 250 mL, with side tube
19 Keck clamps (blue plastic) (2)

NOTE: This equipment is the property of Howard University. Students will be required to pay for damaged or missing items.

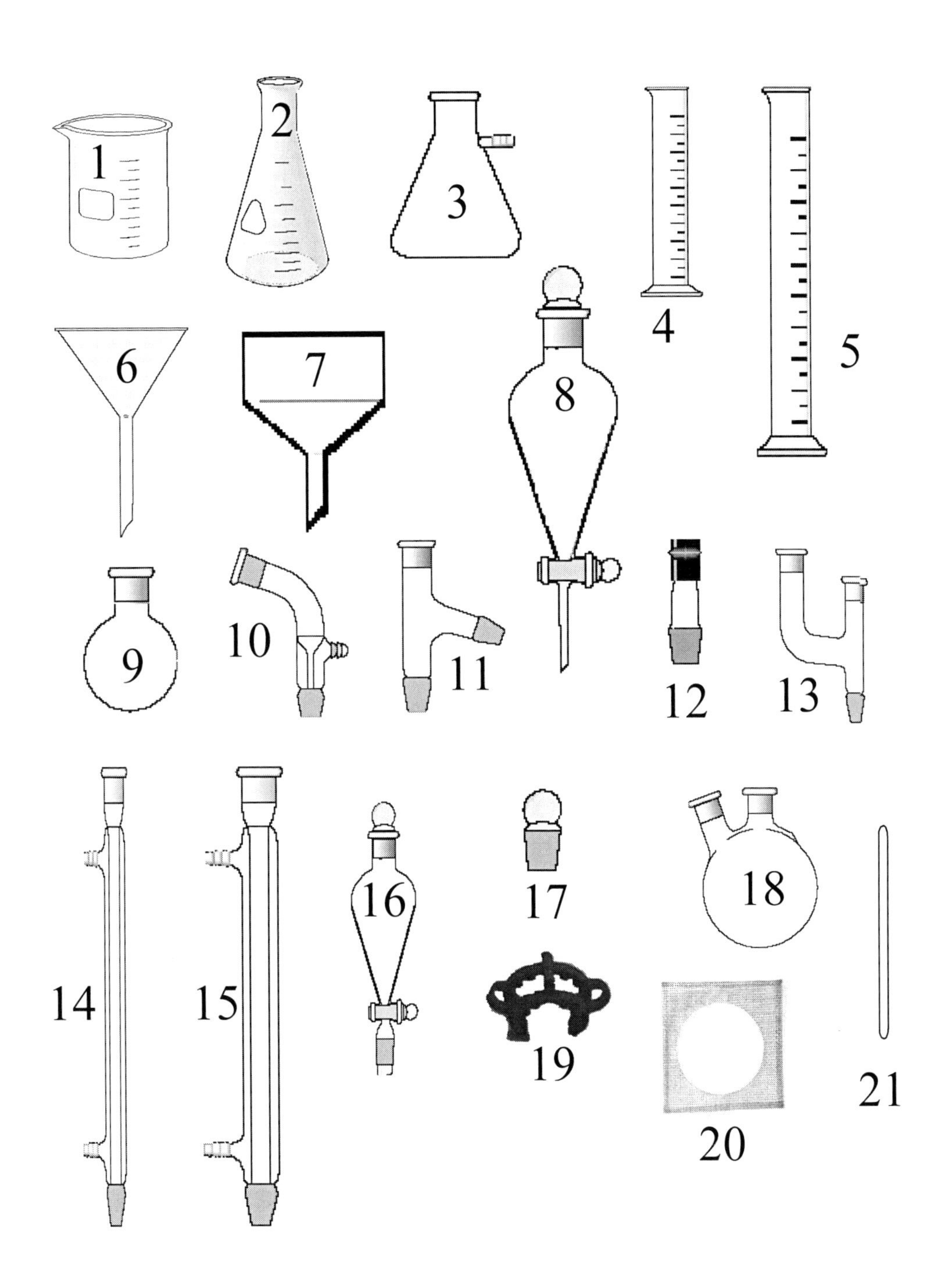
1
2
3
4
5
6
7
8
9
10
11
12
13
14
15
16
17
18
19
20
21

CARE OF GLASSWARE

The best way to clean glassware is by rinsing immediately with water followed by a volatile solvent such as acetone (bp 56 °C), and drying with a paper towel or allowing the glassware to dry. If you have used grease, it is best to remove the grease first: Wipe with a dry paper towel, then with a paper towel moistened with solvent (a nonpolar solvent such as hexane or dichloromethane works best). Then wash as usual.

Glassware which is rinsed with a volatile solvent (such as acetone) will dry very fast. Remember than organic solvent vapors are heavier than air, so lay your flask on its side or clamp it upside down for fastest drying. Compressed air is dirty, noisy, and comparatively ineffective.

If glassware is washed promptly, it is rarely necessary to scrub it with soap and water. Soap (or detergent) and water does not remove stopcock grease very well. Beware of using test tube brushes - they frequently spread grease around rather than remove it, and once contaminated with grease usually stay that way. Paper towels are very effective.

If the stockroom runs out of acetone, other solvents can frequently be used. Or you can get a small bottle of rubbing alcohol (90% isopropyl alcohol is best) from your supermarket or drugstore and use that for cleaning glassware.

Take particular care with **standard taper** glassware, since this is much more expensive than ordinary glassware (such as beakers and Erlenmeyer flasks). Standard taper joints can get stuck (or "frozen") and become very difficult to separate, especially if they are used without proper greasing, left together too long, or exposed to certain caustic materials. Take them apart and clean them when you are finished with an experiment. [Standard taper glassware can often be identified by the numbers (such as 19 or 19/22) on the joints. The first number is the diameter (19 mm) of the joint, the second is the length.]

Most of the "tincture bottles" are <u>not</u> standard taper. Their glass stoppers (which lack the numbers characteristic of most standard taper stoppers) do not fit well. If you use the tincture bottles it is probably better to use corks or rubber stoppers.

Teflon stopcocks are chemically very inert, but rather soft and easily deformed. Loosen them when the stopcock is not in use. If you need to take them apart to clean them, be sure you don't lose any of the little pieces, and put them back together immediately. They do not need any grease. Concerning glass stopcocks: as long as they are well greased and easily turned, there is no need to take them apart.

BE PREPARED

Read the relevant sections of this manual before you come to the laboratory. There will be a quiz every week.

It doesn't hurt to read ahead further - there is always a small possibility that the stockroom will be out of a crucial reagent needed for this week's experiment, but if you are prepared you will be able to skip ahead to the one scheduled for the next week.

For late bulletins, check the chalk boards just outside the laboratory and the whiteboards in both 311 and 312.

LABORATORY NOTEBOOKS

A laboratory notebook is the primary record of experimental work. It should be a record of what you have done and observed in the laboratory, not merely a copy of a procedure from a laboratory manual. It should be complete enough to enable anyone with a basic knowledge of organic laboratory operations to repeat your work.

The notebook must be sturdily bound (not a spiral notebook), preferably about 8 x 10 inches in size, and with ruled pages (rather than graph paper). All entries must be made in permanent ink. If you make any ~~misteaks~~ mistakes, draw a line through your errors so they can still be seen, and write in the correct information.

NEVER REMOVE ANY PAGES FROM THE NOTEBOOK!
DO NOT ERASE ANYTHING FROM THE NOTEBOOK!
NEVER USE ANY CORRECTION FLUID!

The laboratory notebook should be used for laboratory observations and results only. It should not be used for lecture notes or for answers to exercises in the textbook.

Notebook preparation. At the beginning of the semester prepare your notebook as follows:

1. Write your name, course and section number (Chem 145-##), and semester on the cover and inside the front cover (in case any chemicals are spilled on the cover).
2. Number the pages consecutively in the top outside corner. (Use both sides of the paper.)
3. Reserve pages 1-4 for course information and a table of contents, and start the first experiment on page 5.
4. Show your notebook to the instructor for approval.

Pre-lab write-up. Start each experiment on a new page. For each experiment, write down the following information, before you come to the laboratory:

1. A descriptive title.
2. For preparative experiments, a balanced equation, preferably with structural formulas.
3. Any useful information or data, such as molecular weights and physical properties (mp, bp, density, solubility, etc.), for each reagent you expect to use and all expected products. For preparative experiments, this should be in the form of a table of reactants and products. This table should include the weight and the number of moles of each reagent you expect to use. (It is not necessary to

calculate the number of moles of solvents or drying agents.)

4. Determine which reagent is the <u>limiting reagent</u> and calculate the <u>theoretical yield</u> of product.
5. A literature reference to the procedure you are following. (In most cases this will be just the manual page numbers.)
6. In some sections, your instructor may want you to include a flow chart showing the separation and purification process.

If you wish you may include a <u>brief</u> statement of purpose or a reminder of important safety precautions. But do <u>not</u> include a discussion of the reaction mechanism(s), a list of alternative methods of preparation, or anything else that is not directly related to the experimental procedure, unless you are specifically directed to do so.

Experimental Procedure. Bring your notebook to lab every day. When you begin your experiment, enter the date in the margin, then write down what you do and observe.

Write what <u>you</u> do and observe; don't just copy the procedure from the book. What you do will almost certainly be slightly different from what the manual indicates. For example, you may use a different size flask, or decide to measure a liquid by volume instead of weight. Perhaps you find the balance is not as accurate as you thought, and you measure out 4.00 g instead of 4.0000 g. Or you might accidentally weigh out 4.05 g instead of 4.00 g.) Or perhaps the stockroom is out of one reagent and another is substituted.

Write what you <u>did</u>, not what you plan to do. Thus, the notebook will be written in past tense. Never write down something before you actually do it. (If you need a recipe to follow in the laboratory and you don't want to get your manual messy, bring a photocopy of the procedure from the manual.)

It is customary to use passive voice: Instead of "I distilled the compound," write "The compound was distilled."

Try to write or print <u>legibly</u> and <u>accurately</u>. Neatness is also desirable, of course, but it is less important than legibility, accuracy, and completeness. Take particular care with numerical measurements, such as weights or volumes of reagents, reaction times and temperatures, etc.

Be aware of how many significant figures are appropriate. If you weigh out exactly four grams on a balance which reads to the nearest hundredth of a gram, record the weight as 4.00 g, not 4 g or 4.000 g. Fractional quantities should be recorded with an initial zero, <u>e.g.</u>, "0.26 g". (If you record the weight as ".26 g", the decimal point may be overlooked.)

It is convenient to have the quantities of all starting materials and reagents arranged along the left side of the page as shown in the first part of the (fictional)

experimental description shown below. Since you will probably need to recalculate the number of moles of each reagent, it is helpful to leave a space in which you can enter this information later.
Thus, you might initially enter the following:

11.0 ml of n-butanol

Then later, at your convenience, insert the calculated values:

11.0 ml of n-butanol (8.91 g, 0.120 mol)

Note that the measured quantity is always first (followed by the calculated values in parentheses).

The write-up should be complete. A competent organic chemist should be able to duplicate your experiment by reading the procedure in your notebook without having to refer to the manual (or handout). On the other hand, you should not describe how to carry out basic laboratory operations (e.g., how to use a balance or a separatory funnel) which are well known to anyone who has completed an organic laboratory course.

Every sample you prepare should be given a unique sample number which should include your initials and the notebook page number. For example, the second sample on page 23 of PFH's notebook would be given the unique sample number PFH-23-2. The sample number should appear in the notebook, on the label of any sample you turn in, and on any spectra you take of the sample. It should be stressed this is a sample number, not a compound number; if a compound is purified (for example, by distillation, crystallization, or chromatography) it is a new sample and should be given a new sample number.

If your write-up is more than one page, continue on the next page -- don't skip pages. If the next page is occupied, continue on a later page; in that case, be sure to cross-reference on both pages.

Post-lab Write-up. At the conclusion of the experiment, look over your write-up to see that it is complete. Make any necessary calculations, including that of percentage yield (show your calculations). [For the yield calculation, if you used a different quantity of reagent than you originally had planned, be sure to revise your calculation of the theoretical yield as necessary.]

For most preparative-type experiments, it is not usually necessary to add any conclusion, although a statement that might be helpful to someone repeating your experiment would be appropriate. For example: "This reaction was very exothermic. More ice would have been desirable." In contrast, a statement like "In this experiment I learned a lot about crystallization" is not very useful.

References

J. W. Zubrick, "The Organic Chem Lab Survival Manual: A Student's Guide to

Techniques", Wiley, New York, NY, 1984; 2nd ed, 1988; pp 8-20.

H. M. Kanare, "Writing the Laboratory Notebook", American Chemical Society, Washington, DC, 1985.

Preparation of 1-Bromobutane

$CH_3CH_2CH_2CH_2OH$ + NaBr + H_2SO_4 ⟶ $CH_3CH_2CH_2CH_2Br$ + $NaHSO_4$ + H_2O

	$CH_3CH_2CH_2CH_2OH$	NaBr	H_2SO_4	$CH_3CH_2CH_2CH_2Br$
bp	117.7°C			101.6°C
d	0.810		1.84	1.275
MW	74.123	102.894	98.08	137.019
vol	10 ml		12 ml	
			18 M	
wt	(8.1 g)	13.3 g		14.94 g
mol	0.109	0.129	0.216	0.109 moles
	Lim. Reag.			Theor. yield

Theoretical yield = 0.109 mol x 137.019 g/mol = 14.94 g

Reference: Fieser + Williamson, pp 178-180.

WARNING: Sulfuric acid is very corrosive!

1996 Jan 14

To a 100-ml RB flask was added
13.36 g of sodium bromide (crystals) (0.130 mol),
15 ml of water,
and the mixture was swirled until the crystals were dissolved.
To this mixture was added
11 ml of n-BuOH (colorless liquid) (8.91 g, 0.120 mol).
The resulting mixture appeared to contain two liquid layers.

The mixture was cooled in ice for 5 minutes, and
12 ml of conc H_2SO_4 (viscous liquid) (0.216 mol)
was slowly added while swirling in the ice bath.
The mixture appeared to become completely homogeneous, and turned slightly yellow.

After a few minutes in the ice bath, the flask was placed in a heating mantle, a boiling chip was added, a reflux condenser was attached, and the mixture was heated to reflux. (It was smelly - I wished I had done the reaction in a hood.) The mixture became cloudy,

continued on page 23

cont'd from p 19

then two liquid layers formed. The lower layer turned dark yellow.
Heating was continued for ~~45 min~~. 55 min.

The mixture consisted of two liquid layers. The top layer was nearly colorless, the lower layer was yellow-brown. The reflux condenser was removed and the mixture was distilled (apparatus shown on p 66). The crude product was collected in a 25-ml graduated cylinder.

bp 90-116°C

There was obtained 18 ml of crude product as a slightly cloudy, colorless liquid. (The pot residue was a brown liquid, which partially solidified as it cooled.)

The distillate was washed with 10 ml of water, the separatory funnel was cleaned and dried, and the crude product was shaken with 10 ml of cold sulfuric acid and allowed to stand for 10 min before draining off the acid. The organic layer was washed with 10 ml of 10% NaOH solution and dried over anhydrous Na_2SO_4 until the next period. (The stockroom was out of $CaCl_2$.)

1996 Jan 16

A few drops (PFH-23-1) were removed for GC analysis, the remaining liquid was decanted away from the drying agent and fractionally distilled through a column packed with copper turnings. All fractions were colorless liquids.

PFH-23-2 bp 58-98°C 0.21 g (evaporated before GC analysis)

PFH-23-3 bp 98-103°C 9.48 g

In an attempt to get more product, the flask was cooled slightly, the fractionation column was removed, and the remaining liquid was distilled (simple distillation).

PFH-23-4 bp 88-124°C 2.7 g clear, colorless liquid

(mostly bp 115-120°C)

About 3 ml of a brown liquid remained in the distillation flask. This residue was discarded.

From the boiling temperatures it appears that PFH-23-3 is the desired 1-bromobutane and that PFH-23-4 is largely unreacted butanol.

Yield: $\frac{(9.48/137.019)}{0.120} = 0.5766 = 58\%$

1996 Jan 18

GC analysis. Samples were dissolved in CH_2Cl_2 (1 drop of sample in 2 mL CH_2Cl_2) and analyzed (1 μL of sample solution used) on GC "C" using a HP-1 silicone column. Conditions:
initial temp. 40°C (2 min), program rate 40°/min, final temp. 160°C (0 min), injection Temp: 220 °C, detector temp: 250 °C, FID, He carrier gas, head pressure: 2 psi, attenuation: 5, chart speed: 2.

Posted retention times:

1-bromobutane	1.13 min
1-butanol	0.68 min
dibutyl ether	3.72 min

Analysis of GC's (see attached GC's):

Solvent CH_2Cl_2 had RT's of: 0.26. 0.47, 2.18, 2.31, 2.94, 3.10, and 4.25 min.

Sample PFH 23-1:

$$\%\ 1\text{-}butanol = 100 \cdot \frac{145270}{388+1412+332+145270+707+1494154+3228+1685+458+9464+727+1594+3783+74776}$$

$$= 100 \cdot \frac{145270}{1737978} = 8.4\ \%$$

$$\%\ 1\text{-}bromobutane = 100 \cdot \frac{1494154}{1737978} = 86.0\ \%$$

$$\%\ dibutyl\ ether = 100 \cdot \frac{74776}{1737978} = 4.3\ \%$$

also contains 11 other unidentified minor components not present in CH_2Cl_2

Sample PFH 23-3:

$$\% \text{ 1-}butanol = 100 \cdot \frac{202491}{355+1898+202491+787+1639094+2812+1764+498+10474+807+1432+4162+104971}$$

$$= 100 \cdot \frac{202491}{1971545} = 10.3 \ \%$$

$$\% \text{ 1-}bromobutane = 100 \cdot \frac{1639094}{1971545} = 83.1 \ \%$$

$$\% \ dibutyl \ ether = 100 \cdot \frac{104971}{1971545} = 5.3 \ \%$$

also contains 10 other unidentified minor components not present in CH_2Cl_2.

Sample PFH 23-4:

$$\% \text{ 1-}butanol = 100 \cdot \frac{1616185}{234+1416+1616185+971937+2107+1240+357+6643+480+1103+2546+219455}$$

$$= 100 \cdot \frac{1616185}{2823703} = 57.2 \ \%$$

$$\% \text{ 1-}bromobutane = 100 \cdot \frac{971937}{2823703} = 34.4 \ \%$$

$$\% \ dibutyl \ ether = 100 \cdot \frac{219455}{2823703} = 7.8 \ \%$$

also contains 9 other unidentified minor components not present in CH_2Cl_2.

The IR spectrum (liquid film) of PFH-23-3 was obtained. Peaks at 3356, 2958, 2927, 2856, 1465, and 1378 cm^{-1} were observed. Assignments: peaks at , 2958, 2927, 2856 cm^{-1} are the methyl group (CH_3) and methylene group (CH_2) stretching bands; peaks at 1465, and 1378 cm^{-1} are the methyl group (CH_3) and methylene group (CH_2) bending bands; broad peak at 3356 cm^{-1} is an OH stretching band either from the 1-butanol in the sample or from H_2O.

* RUN # 7 JAN 1, 1901 01:44:08 1/18/96 CH_2Cl_2 alone

START: not ready

0.256

0.467

2.179

2.308

2.943

3.101

4.253

STOP

RUN# 7 JAN 1, 1901 01:44:08

AREA%

RT	AREA	TYPE	WIDTH	AREA%
.256	76193280	SHB	.016	99.98730
.467	1515	TBB	.028	.00199
2.179	591	PV	.060	.00078
2.308	1278	VV	.048	.00168
2.943	298	PV	.039	.00039
3.101	948	VP	.047	.00124
4.253	5094	VV	.127	.00668

TOTAL AREA=7.6203E+07
MUL FACTOR=1.0000E+00

* RUN # 9 JAN 1, 1981 02:04:52 1/18/96 PFH-23-1

START

0.115
0.259
0.340
0.476
0.624
0.691
0.852
1.143
1.879
2.314
2.363
2.540
2.697
2.788
2.946
3.024
3.103
3.280
3.720
4.245

STOP

RUN# 9 JAN 1, 1981 02:04:52

AREA%

RT	AREA	TYPE	WIDTH	AREA%
.115	388	VB	.038	.00050
.259	75890304	SPB	.017	97.75178
.340	1412	TBB	.017	.00182
.476	1717	TBB	.029	.00221
.624	332	TBV	.025	.00043
.691	145270	TVB	.033	.18712
.852	707	BP	.056	.00091
1.143	1494154	PB	.051	1.92457
1.879	3228	PV	.056	.00416
2.314	1100	VV	.043	.00142
2.363	1685	VV	.052	.00217
2.540	458	VV	.041	.00059
2.697	9464	VV	.036	.01219
2.788	727	VV	.038	.00094
2.946	455	VV	.056	.00059
3.024	1594	VV	.041	.00205
3.103	1024	VV	.055	.00132
3.280	3783	VP	.030	.00487
3.720	74776	PB	.027	.09632
4.245	3282	BV	.087	.00423

TOTAL AREA=7.7636E+07
MUL FACTOR=1.0000E+00

* RUN # 8 JAN 1, 1901 01:54:52 1/18/96 PFH-23-3

START

0.115

0.255

0.335

0.466

0.677

0.835

1.128

1.869

2.308
2.356

2.536

2.694

2.778

3.021
3.099

3.280

3.721

4.250

STOP

RUN# 8 JAN 1, 1901 01:54:52

AREA%

RT	AREA	TYPE	WIDTH	AREA%
.115	355	PB	.029	.00050
.255	69351680	SPB	.016	97.22701
.335	1898	TBP	.017	.00266
.466	1687	TBB	.029	.00237
.677	202491	PB	.033	.28388
.835	787	BP	.055	.00110
1.128	1639094	PB	.051	2.29792
1.869	2818	PB	.048	.00395
2.308	1010	VV	.043	.00142
2.356	1764	VV	.052	.00247
2.536	498	VV	.041	.00070
2.694	10474	VV	.036	.01468
2.778	807	VB	.036	.00113
3.021	1432	VV	.036	.00201
3.099	699	VV	.046	.00098
3.280	4162	VP	.030	.00583
3.721	104971	PB	.028	.14716
4.250	3115	BV	.091	.00437

TOTAL AREA=7.1330E+07
MUL FACTOR=1.0000E+00

* RUN # 11 JAN 1, 1981 02:26:11 1/18/96 PFH 23-4

START

0.117

0.255

0.464

0.686

1.125

1.859

2.305

2.534

2.692

2.777

3.020

3.099

3.277

3.720

4.239

STOP

RUN# 11 JAN 1, 1981 02:26:11

AREA%

RT	AREA	TYPE	WIDTH	AREA%
.117	234	BV	.026	.00036
.255	62928576	SHB	.017	95.69594
.464	974	TBV	.024	.00148
.514	1416	TVP	.030	.00215
.686	1616185	TPV	.034	2.45774
1.125	971937	PB	.050	1.47803
1.859	2107	PV	.060	.00320
2.305	1009	VV	.047	.00153
2.355	1240	VV	.055	.00189
2.534	357	PV	.041	.00054
2.692	6643	VV	.036	.01010
2.777	480	VV	.035	.00073
3.020	1103	VV	.042	.00168
3.099	793	VV	.051	.00121
3.277	2546	VP	.030	.00387
3.720	219455	PB	.027	.33373
4.239	3873	VV	.109	.00589

TOTAL AREA=6.5759E+07
MUL FACTOR=1.0000E+00

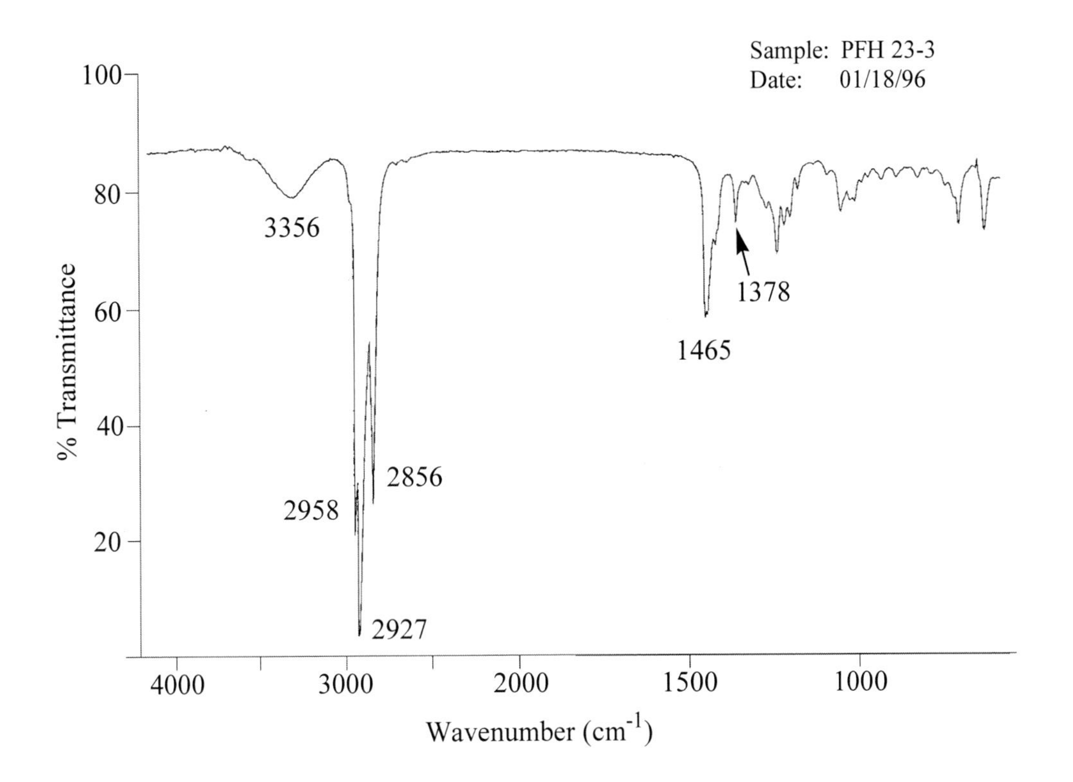
Sample: PFH 23-3
Date: 01/18/96
100
80
60
40
20
% Transmittance
3356
2958
2927
2856
1465
1378
4000
3000
2000
1500
1000
Wavenumber (cm-1)

FORMAL REPORTS

A formal report is somewhat like a scientific article in a research journal. It should present your results clearly, and relate them to the existing knowledge in the literature. The report should consist of the following parts:

Title. This should be brief but descriptive. Follow with your name, course number, and date.

Abstract. A concise (only several sentences) and informative summary of what you did and the results you obtained.

Introduction. Provide enough background information to put the experiment into perspective, but use your own words rather than copying directly from the textbook. Provide a literature reference for every statement which is based on your textbook reading or library research. (The best reports will include several references from the library.) Tell the reader what experiment was carried out (and perhaps why), but do not reveal your results just yet. Include a balanced equation using structures, if appropriate. Be sure you provide a reference to the procedure you are following.

Experimental Section. This is the most important section of your report. In some ways it is similar to the write-up in your notebook, but it should be better written, more concise, and follow a fairly rigid format. It should be written in passive voice, and it must include all necessary data and results, including the following:

- quantities of all starting materials (give the measured values first, any calculated values in parentheses);
- number of moles of all reagents (but not solvents or drying agents);
- reaction times and temperatures;
- workup and purification procedure;
- quantity of product, and % yield;
- observed properties of the product (clarity, color, unusual odor, mp, bp, GC results, IR spectra, etc.);
- comparison of mp, bp, or spectra with literature values.

Always record melting points and boiling points as ranges. If the literature mp or bp is known, compare your result to the literature value. Always give the weight of the product, but do not say "the product was weighed".

Do not include any data or observations that are not recorded in your notebook or taken directly from a spectrum. Do not show a table of reagents or extensive calculations -

these are more appropriate for the laboratory notebook. Do include a brief calculation of percentage yield only. Do not include any conclusions you may draw from the data - that belongs in the next section.

Results and Discussion. Briefly summarize your results, and draw any conclusions that you think are appropriate. You have considerable freedom here. You may include a discussion of reaction mechanisms, important side reactions, safety precautions, or your purification scheme. You may offer possible explanations for low yields or impure products, or suggestions on how the experiment might be improved. Be sure to account for any unusual observations mentioned in the Experimental Section. Include a discussion of any instrumental analysis such as GC or IR. Do not present any new data or information which does not appear in the Experimental Section. As before, provide a literature reference for any statement which is not based on your own observations.

References and Notes. Any statement you have made in the preceding sections which is not based on your own observations or knowledge must be documented. This applies to any data taken from this manual or any other printed source, as well as any information from the instructor or from lecture notes. Collect your references here, numbered in the order they are cited in your report. If you use a reference more than once, use the same number again.

For each citation, provide names of all authors. Make sure the names are spelled correctly. Note that only last names (e.g. Moore and Dalrymple) are used if you mention the authors' names in the text, but that the first initials are included in the References and Notes.

References to books should include the title of the book in quotation marks, the publisher, city, and date. (The two books listed below may be helpful to those learning to write scientific reports.

J. S. Dodd, Ed., "The ACS Style Guide: A Manual for Authors and Editors", 2nd edition, American Chemical Society, Washington, D. C., 1997, ISBN: 0-8412-3462-0.

W. Strunk, Jr., and E. B. White, "The Elements of Style", 3rd ed, Macmillan, New York, 1979.

References to journal articles should include the underlined journal abbreviation, the volume, year, and complete pages. Two acceptable formats are shown below. In the first, currently used by ACS journals, the initials follow the names, and the journal abbreviation is followed by year, volume, and inclusive pages. In the second format, formerly used by many ACS journals and still used by some other journals (e.g., Science), the initials come first, and the journal abbreviation is followed by volume, pages, and year. With either format, the article title may be included. Choose one of these formats and follow it consistently.

Katsuki, T.; Sharpless, K. B. J. Am. Chem. Soc. **1980**, 102, 5974-5976.
G. Stork, P. Rosen, N. Goldman, R. V. Coombs, and J. Tsuji, J. Am. Chem. Soc., **87**, 275-286 (1965).

It should not be necessary to refer to lectures and personal communications; where appropriate, see the following example:

A. B. Smith, lecture, Howard University, September 12, 1993.

Report Format

The report should be word processed, double spaced, with 1" margins, on one side of the paper, and stapled in the upper left corner only. The output from a high-quality computer printer is ideal. All drawings and structures must be in ink. Handwriting (in black or blue ink) is acceptable only if it is extremely neat and legible. Do not use a cover page. (Plastic covers or folders will be discarded.)

Your instructor will tell you whether to use the original or xeroxed copies of GC charts and IR spectra. Attach all GC charts and IR spectra. GC's should be labeled with your name, the date, and the GC conditions (temperature, time, program rate); all major peaks should be identified (name or structural formula). IR spectra should be labeled with your name, the date, the name (or structural formula) of the compound, and the phase (liquid film, KBr pellet, or nujol mull).

General Comments

1. Never start a sentence with a number. Reword the sentence instead:
   To a 100-ml flask was added 6.5 g of compound ...
   A mixture of 6.5 g of compound and ...
   A 6.5-g sample of compound was ...
2. Always use numbers (numerals) for units of measurement, such as quantities of reagents and anything that might be fractional:
   6 ml   0.30 g   3.5 hr   133°C   0.68 ml
   On the other hand, for countable items, spell out the number:
   six beakers   three 25-ml portions
3. Be careful with significant figures. When you weigh out a reagent, you have to decide how many decimal places are significant (e.g., 5.0 g, 5.00 g, or 5.000 g, etc.). If you start with 5.0 g of the limiting reagent (two significant figures), it is meaningless to calculate the final yield to four decimal places.

4. Concerning clear and colorless. "Clear" means not milky; "colorless" means water-white. Water is clear and colorless. Apple juice and tea are clear, but not colorless. Milk is colorless, but not clear. Is that clear?

Study the sample reports attached. It may also be helpful to look at a few articles in recent issues of chemical journals such as the Journal of the American Chemical Society or the Journal of Organic Chemistry.

[SAMPLE FORMAL REPORT]

(Note: Abstract, Introduction, Experimental Section, and Discussion are **double spaced**)

THE PREPARATION OF CYCLOHEXYL CHLORIDE FROM CYCLOHEXANOL

Howard U. Student

Chemistry 145-01

2009 June 30

Abstract. Cyclohexyl chloride was prepared in 46% yield from cyclohexanol and concentrated hydrochloric acid.

Introduction

Alcohols can be converted to alkyl halides by a number of reagents, including thionyl chloride, phosphorus halides, and hydrohalic acids.[1,2] Toward hydrochloric acid, tertiary alcohols are generally more reactive than secondary alcohols, which are more reactive than primary alcohols.[2,3] Cyclohexanol is reported to be exceptional in its reactivity toward HCl,[1] and can be converted to its chloride more easily than other secondary alcohols. In this experiment, cyclohexanol was converted to cyclohexyl chloride (chlorocyclohexane) by treatment with concentrated hydrochloric acid, using the procedure of Moore and Dalrymple.[1]

OH HCl Cl

Experimental Section

Chlorocyclohexane. A mixture of 20.2 ml (19.4 g, 0.194 mol) of cyclohexanol and 80 ml (0.96 mol) of concentrated hydrochloric acid were heated at reflux for 2.5 hr. The mixture was cooled and the layers were separated. The organic layer was washed successively with 10 ml of concentrated sulfuric acid, 10 ml of water, and two 20-ml portions of 10% Na_2CO_3 solution, then dried ($CaCl_2$) and distilled, yielding 10.7 g (46% yield)[4] of chlorocyclohexane as a clear, pale yellow liquid, bp 139-145°C (lit.[1] bp 138-142°C); IR (film) 3250, 2930, 2850, 1462, 1260, 995 cm^{-1}. The infrared spectrum was comparable to that reported by Roberts, et al.,[5] except for the small, broad absorption at 3250 cm^{-1}.

Gas chromatographic analysis on a HP-1 column at 60°C showed the major peak at 3.8 min (94%), and a small peak at 5.7 min (5%). Under these conditions, the retention time of cyclohexanol was 5.7 min.

Reaction with 2,4-DNP Reagent.[6] Following the procedure of Roberts, et al.,[7] to 2 ml of the 2,4-dinitrophenylhydrazine reagent solution was added two drops of the chlorocyclohexane prepared above. A copious yellow precipitate appeared; it was recrystallized from ethanol, giving 0.02 g of yellow crystals, mp 125-127°C.

Results and Discussion

Cyclohexanol was converted to chlorocyclohexane in 47% yield. Analysis by gas chromatography showed the compound was obtained in 94% purity, with the major impurity (5%) having the same retention time as cyclohexanol. The infrared spectrum showed very strong peaks at 2800-2980 cm^{-1}, corresponding to the expected aliphatic C-H stretching vibrations.[8] There was a small peak at 3250 cm^{-1} (corresponding to hydroxyl),[8] consistent with the small amount of cyclohexanol impurity or a small amount of water. The absence of peaks at 3010-3080 cm^{-1} (corresponding to olefinic C-H)[8] indicates that no dehydration to form cyclohexene took place. The procedure was easy to follow and with one exception the experiment was routine. In a misguided attempt to prepare a 2,4-dinitrophenylhydrazone derivative, a yellow precipitate melting at 125-127°C was obtained. In retrospect, there should not have been any reaction. Perhaps the glassware was wet with traces of acetone; this would have resulted in the 2,4-DNP derivative of acetone, which is known to have a mp of 128°C.[9]

References and Notes[10]

1. J. A. Moore and D. L. Dalrymple, "Experimental Methods in Organic Chemistry", 2nd ed, Saunders, Philadelphia, 1976, pp 152-154.

2. S. G. Wilkinson, in "Comprehensive Organic Chemistry", D. Barton and W. D. Ollis,

Eds., Volume 1, J. F. Stoddart, Ed., Pergamon Press, Oxford, 1979; pp 633-638.

3. T. W. G. Solomons, "Organic Chemistry", Wiley, New York, N. Y., 1976, p 602.

4. % Yield: $\frac{(10.7/118.5)}{(19.4/100)} = \frac{(0.0903 \text{ mol})}{(0.194 \text{ mol})} = 46.5\%$

5. R. M. Roberts, J. C. Gilbert, L. B. Rodewald, and A. S. Wingrove, "Modern Experimental Organic Chemistry", 4th ed, Saunders, Philadelphia, 1985, p 329.

6. Nobody in his or her right mind would expect chlorocyclohexane to react with 2,4-DNP reagent; this paragraph was merely added to show how to include a second reaction in the experimental section.

7. Ref 5, pp 700-701.

8. Ref 5, pp 210-212.

9. Ref 5, p 756.

10. Note that when a reference is used more than once, the full details need not be repeated (see refs 7-9).

[Your instructor will indicate whether the original or a copy of the GC tracing(s) and IR spectra should be attached. All GC's should be labeled with your name, the date, and the GC conditions, and each significant peak should be identified. IR spectra should be labeled with your name, the name of the compound, and the phase (e.g., liquid film).]

SHORT REPORTS

Your instructor may ask you to write up some experiments as short reports and others as formal reports.

Short reports should include the following:

Your name
Brief title
Balanced equation (using structural formulas)
Quantities of all starting reagents
    either weight (moles) or volume (weight, moles)
    pay attention to significant digits
Reaction time and temperature
Separation procedures:
    Solvent extraction, distillation, crystallization
Product: weight, mp or bp
Calculation of % yield
IR peaks (and assignments)
GC analysis (table)
Attach the original/copy GC and IR

Formal reports should contain all of the above information, plus a review of the pertinent literature, a complete experimental section, and a brief summary or conclusion.

[SAMPLE SHORT REPORT]

Howard U. Student
Chem 145-03
1996 Feb 9

Chlorocyclohexane
Notebook pages 25-26

$$\text{cyclohexanol (C}_6\text{H}_{11}\text{OH)} + HCl \longrightarrow \text{chlorocyclohexane (C}_6\text{H}_{11}\text{Cl)} + H_2O$$

Cyclohexanol was converted to chlorocyclohexane by reaction with concentrated HCl.

Reagents 20.2 ml (19.4 g, 0.194 mol) of cyclohexanol
80 ml (0.96 mol) of concentrated HCl

Reaction conditions 2.5 hours at reflux

Product isolation Product washed with H_2SO_4, water, aqueous Na_2CO_3, then dried ($CaCl_2$), and distilled.

Product 10.7 g (0.0903 mol, 46% yield) of a clear, pale yellow liquid bp 139-145°C

Calculation of % yield

$$\frac{(10.7/118.5)}{(19.4/100)} = \frac{(0.0903\ \text{mol})}{(0.194\ \text{mol})} = 46.5\%$$

IR spectrum:

3250 cm^{-1}	O-H stretch (from cyclohexanol or water)
2930	sp^3 C-H stretch
2850	sp^3 C-H stretch
1462	C-H bend
1260	
995	

GC analysis: 3.8 min (94%) chlorocyclohexane
5.7 min (5%) cyclohexanol

[Attach original/copies of GC and IR spectra. All GC's should be labeled with your name, the date, and the GC conditions, and each significant peak should be identified. IR spectra should be labeled with your name, the name of the compound, and the phase (e.g., liquid film).]

SIGNIFICANT FIGURES

Suppose your procedure calls for 20 g of cyclohexanol (MW 100.161). Using a pocket calculator, it's easy to calculate the number of moles: 0.19967852 moles. How many of these digits are meaningful? It depends on what balance you use, and how carefully you weigh out the cyclohexanol.

Many of the balances in the organic lab can read to the nearest 0.01 g. If you just weigh out 20 g without bothering to record anything to the right of the decimal point, your weight is only good to two significant digits, and it is not meaningful to calculate the number of moles to more than two significant figures (0.20 moles).

$$\frac{20}{100.161} = 0.19967852 \quad \textit{rounding off gives } 0.20$$

Suppose we weigh out 20.35 g of cyclohexanol. This is four significant digits, and if we write down 20 g (only two digits) in the notebook we are losing information. Any calculations we make (such as calculating the number of moles) should be done to the same degree of precision. Thus, the correct answer is 0.2032 mol (four significant digits), or 203.2 mmol (millimoles).

$$\frac{20.35}{100.161} = 0.20317289 \quad \textit{rounding off gives } 0.2032$$

On the other hand, if you have carefully weighed your compound to the nearest 0.01 g, but you use an estimated MW of 100 (only three significant digits), then your answer is only good to three digits.

Sometimes it is easier to measure volume instead of weight, and use the published value for density to calculate the weight. However, it is difficult to read volumes to more than three significant digits, and densities are often given with only two digits. Thus, 20.9 ml x 0.96 g/ml = 20.064 g, but this number is not accurate to more than three digits.

Don't be in too much of a hurry to round off your calculations. Often a number is used in further calculations. For example, the number of moles is usually used for the calculation of percentage yield. When this is the case, keep one or two extra decimal places (i.e. five digits instead of four) in the intermediate values, and round off to four significant digits in the final result. (Percentage yields are never calculated to more than three digits.)

Zeros on the left are not counted. Thus, a number like 0.00123 has three significant digits. Zeros on the right may be significant. This is one reason for using scientific notation.

For example, the distance between the earth and the sun is about 93 million miles. If it is written as 93,000,000 miles, it gives the misleading impression that there are eight significant figures. Better to write it like this: 9.3×10^7 miles. This makes it clear that only two digits are significant. If it were written as 9.300×10^7 miles, this would indicate four significant figures.

So don't just copy the numbers from your pocket calculator - think about how many decimal places are meaningful.

APPARATUS AND TECHNIQUES

STANDARD-TAPER GLASSWARE

Much of the glassware used in the Organic Chemistry laboratory has a special type of joint called a Standard-Taper Ground Glass joint. This type of joint allows two pieces of glassware having the same sized joint to be joined together without requiring corks, rubber stoppers, or glass tubing. Standard-taper glassware is designated by the symbol $\$$ followed by two numbers separated by a slash, i.e. 19/22. The first number of the pair is the diameter in mm of the widest part of the joint and the second number is the length in mm of the joint. All the standard-taper glassware we use is $\$$ 19/22. There are two types of joints in standard-taper glassware, male and female. Male standard-taper joints have the ground glass surface on the exterior of the joint, while female standard-taper joints have the ground glass surface on the interior of the joint. Any male standard-taper joint will fit into any female standard-taper joint of the same size.

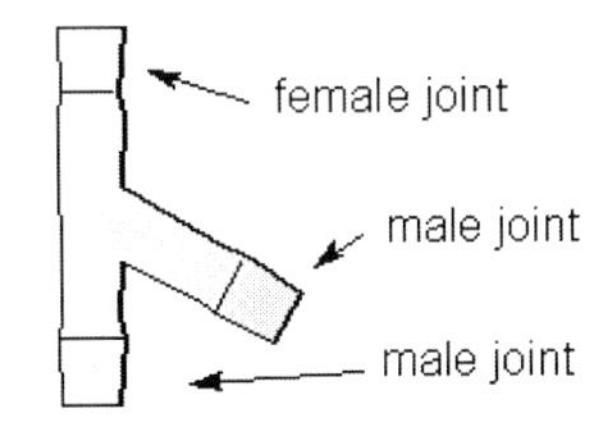

Standard-taper joints must be lubricated with grease so they do not freeze and become impossible to separate. Only a very thin film of grease is required for the joint to be properly lubricated. To lubricate a standard-taper joint, first place a very, very small dab of stopcock grease on the upper outside (ground glass surface) part of the male joint. Then insert the greased male joint into the female joint and gently rotate to spread the grease throughout the joint. A properly lubricated ground glass joint will appear uniformly translucent rather than cloudy. If the grease oozes out of the joint, you used too much.

Standard-taper joints that are not lubricated at all or that are left together for long periods can become frozen. When you have completed the experiment and your apparatus has cooled, take the joints apart promptly. Wipe the grease off with a paper towel or tissue. If you can not easily separate the joint, ask your instructor for help.

MELTING POINTS

The temperature at which a solid is in equilibrium with its corresponding liquid is the melting point. The temperature is generally a small range, rather than a single temperature, and is reported as mp 103-105 °C.

The melting point is a physical constant which is useful in identifying an organic compound. In the synthesis of a compound which has already been prepared and characterized, the melting point of your product should agree with the value previously reported (literature value). Recording the melting point is routinely part of all experiments in which the reaction product is a solid.

Melting point is also a criterion of purity. When solids contain impurities, their melting points are lower. (Melting point depressions can be used to roughly determine the molecular weights of compounds.) The range of the melting point is broader when impurities are present. Thus, a pure compound may give mp 103-103.5°C, and the same compound with impurities present may show mp 97-101°C.

Because the melting point characterizes a compound, an unknown compound may tentatively be identified from its melting point. However, many compounds have the same melting point. In order to use mp to establish identity conclusively, the unknown compound is mixed with the compound which has the same mp. If the two compounds are identical, the mp of the mixture will be the same as that of the individual compounds. If the unknown is different from the known compound, the mp of the mixture will almost always be lower than that of the individual compounds.

Capillary tube

Sample 3-4 mm

Melting points of solid organic compounds will be determined using a Mel-Temp II apparatus (Laboratory Devices). A capillary tube containing the sample is placed in the apparatus (up to 3 tubes can be inserted at the same time) and the sample is observed through a magnifying lens. The rate of heating is controlled by the setting of the Power Level knob. The temperature is read from a thermometer that is also inserted into the apparatus. For optimal results the capillary tube should contain only 3 to 4 mm of sample (see diagram) and the rate of increase in temperature should be kept to about 1 °C per minute. Too much sample results in uneven heating, so that some of the sample may be liquid and some may be solid because the heat has not reached it, and the observed mp will be incorrect.

DRYING AGENTS

Before an organic compound is isolated in pure form by either crystallization or distillation, any water present must be removed. Water can be removed using a two step process. First, the organic solution is extracted with a saturated sodium chloride solution. Water is more soluble in the highly ionic, saturated sodium chloride solution than in an organic solvent. Then the crudely dried organic solution is treated with a drying agent. A drying agent is a compound that can absorb water. The most useful drying agents are solids that can quickly incorporate water molecules into their crystal structures, i.e. compounds that undergo solid phase hydration reactions. Examples of this type of drying agent are:

Calcium Chloride	$CaCl_2 + 6\ H_2O$	$\rightleftarrows$	$CaCl_2 \cdot 6\ H_2O$
Magnesium Sulfate	$MgSO_4 + 7\ H_2O$	$\rightleftarrows$	$MgSO_4 \cdot 7\ H_2O$
Sodium Sulfate	$Na_2SO_4 + 10\ H_2O$	$\rightleftarrows$	$Na_2SO_4 \cdot 10\ H_2O$

You will use anhydrous (contains no water) sodium sulfate as a drying agent in many of the experiments you perform. Anhydrous sodium sulfate has a high capacity (reacts or absorbs a large number of water molecules), is inexpensive, and comes in a granular form so it can be separated from the dried organic solution by decantation (carefully pouring off the liquid). Because the absorption of water by these drying agents is reversible and dependant on temperature, anhydrous sodium sulfate can only be used as a drying agent at or below room temperature. With anhydrous sodium sulfate 5 to 10 minutes is required for equilibrium to be achieved.

Procedure: To the organic layer contained in an Erlenmeyer flask (the flask should be no more than half full) add 4 to 5 g of anhydrous sodium sulfate (enough to just cover the bottom of the flask). Cork the flask. Gently swirl the flask occasionally over 5 to 10 minutes. If all the sodium sulfate appears clumped together, add some more anhydrous sodium sulfate, and continue to periodically swirl the flask for an additional 5 to 10 minutes. The dried solution can then be decanted (carefully poured off) from the drying agent into a dry flask for crystallization or distillation. The drying agent remaining in the Erlenmeyer flask can be rinsed with a few mL of the solvent, and this liquid can then also be decanted into the flask to be used for crystallization or distillation.

SIMPLE DISTILLATION

Simple distillation can be used to separate a liquid from other materials that are not volatile, or to separate two liquids that have boiling points that differ significantly (more than 50-60 °C). Essentially the liquid is boiled and the vapors produced are condensed and collected. The apparatus used for a simple distillation is shown in the Figure labeled "Simple Distillation". The required equipment includes a distilling flask (usually a round bottomed flask that has a volume about twice that of the liquid to be distilled), a connecting adapter, a West condenser, a vacuum adapter, an outlet adapter with a thermometer inserted through the neoprene fitment, and a receiver (generally a graduated cylinder or a test tube). A heating mantle sized to fit the distilling flask is used as the heat source. Assemble the apparatus for a simple distillation as follows:

1. Clamp the distilling flask high enough above the bench top (at least 6") so that the heating mantle and its supporting ring can be lowered completely below the distilling flask. Use the vertical support rod nearest to the Powermite electrical connector.
2. Add one or two boiling chips to the distilling flask and then add the liquid to be distilled to the distilling flask using a funnel.
3. Insert the connecting adapter into the distilling flask, making sure that the ground glass joint is properly greased. You do NOT need to use a clamp to hold the connecting adapter. Next attach the West condenser to the connecting adapter using a Keck clamp, again making sure that the ground glass joint is properly greased. Also use a clamp to support the West condenser. Then attach the vacuum adapter to the lower end of the West condenser using a Keck clamp.
4. Insert the thermometer assembly (thermometer inserted through the neoprene fitment attached to the outlet adapter. Be sure the thermometer is positioned properly (see the Figure; the top of the bulb of the thermometer should be even with the bottom of the side arm joint of the connecting adapter).
5. Position the receiver under the bottom outlet of the vacuum adapter, by either supporting it with a ring and wire gauze or a clamp.
6. Connect rubber hoses (2) to the West condenser. Be sure that the hose from the water source (bottom of an aspirator) is connected to the lower connector on the West condenser. The hose from the upper connector on the West condenser should lead to the drain.
7. Use a ring clamp to support a heating mantle that exactly fits the distilling flask. The power cord from the heating mantle MUST be plugged into the Powermite outlet.
8. Before proceeding with the distillation have your instructor check your

completed apparatus.

9. Turn on the water to the West condenser. A gentle flow is sufficient. An excessive water flow rate can disconnect a hose and cause a flood.
10. To start the distillation, turn the knob on the Powermite to a setting of 35-40. An indicator light on the Powermite unit should turn on.

Simple Distillation

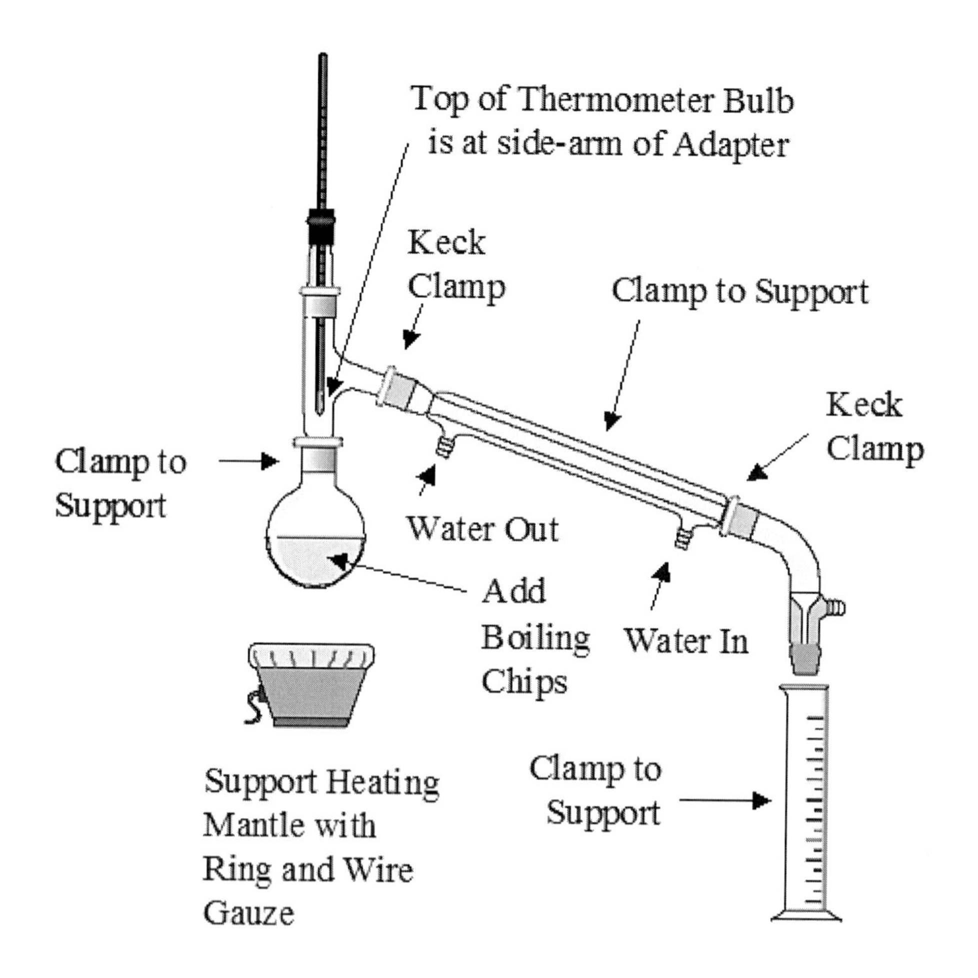

The boiling point of the distillate is read from the thermometer when liquid drips from the bulb. The distillation must be stopped when only a few mL of solution remain in the distilling flask. NEVER distill a liquid to dryness. After the distillation is completed, turn off the Powermite, lower the heating mantle completely away from the distilling flask, and allow the distilling flask and connecting adapter to cool to room temperature.

FRACTIONAL DISTILLATION

Fractional distillation is used to separate liquids that have boiling points that are relatively close together. The apparatus described below should be able to separate liquids that have boiling points that differ by as little as 20-30 °C. In a simple distillation a liquid is boiled (vaporized) and then condensed and collected. In a fractional distillation many vaporization/condensation cycles are carried out before the final condensate is collected. These vaporization/condensation cycles occur on the surface of the packing material in a fractionating column. In this course, copper mesh will be used as the packing material in the fractionating column. The apparatus for a fractional distillation is essentially the same as that for a simple distillation except that a fractionating column is placed between the distilling flask and the connecting adapter (see the Figure labeled "Fractional Distillation"). The required equipment includes a distilling flask (usually a round bottomed flask that has a volume two times the volume of the liquid to be distilled), a fractionating column, a connecting adapter, a West condenser, a vacuum adapter, an outlet adapter with a thermometer inserted through the neoprene fitment, and a receiver (generally a graduated cylinder or a test tube). A heating mantle sized to fit the distilling flask is used as the heat source. Assemble the apparatus for a fractional distillation as follows:

Obtain some copper mesh from the stockroom if a packing material will be used in the fractionating column.

1. Using your glass stirring rod gently insert the copper mesh in the top (female end) of the fractionating column so that it is loosely but evenly distributed along the length of the column. The fractionating column is wider than the West condenser and has glass indentations near the male end.
2. Clamp the distilling flask high enough above the bench top (at least 6") so that the heating mantle and its supporting ring can be lowered completely below the distilling flask. Use the vertical support rod nearest to the Powermite electrical connector.
3. Add one or two boiling chips to the distilling flask and then add the liquid to be distilled to the distilling flask using a funnel.
4. Insert the packed fractionating column into the distilling flask making sure

that the ground glass joint is properly greased. Use a clamp to support the distillation column.

5. Insert the connecting adapter into the top of the fractionating column, making sure that the ground glass joint is properly greased. You do NOT

Fractional Distillation

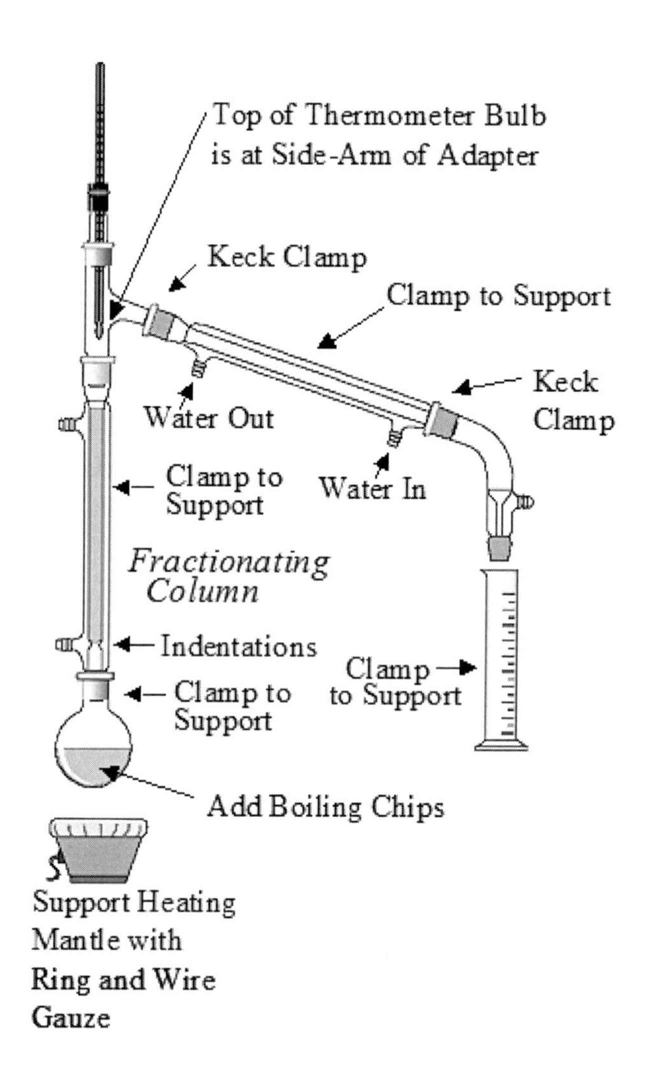

need to use a clamp to hold the connecting adapter. Next attach the West condenser to the connecting adapter using a Keck clamp, again making sure that the ground glass joint is properly greased. Also use a clamp to support the West condenser. Then attach the vacuum adapter to the lower end of the West condenser using a Keck clamp.

6. Insert the thermometer assembly (thermometer inserted through the neoprene fitment attached to the outlet adapter. Be sure the thermometer is positioned properly (see the Figure; the top of the bulb of the thermometer should be even with the bottom of the side arm joint of the connecting adapter).
7. Position the receiver under the bottom outlet of the vacuum adapter, by either supporting it with a ring and wire gauze or a clamp.
8. Connect rubber hoses (2) to the West condenser. Be sure that the hose from the water source (bottom of an aspirator) is connected to the lower connector on the West condenser. The hose from the upper connector on the West condenser should lead to the drain.
9. Use a ring clamp to support a heating mantle that exactly fits the distilling flask. The power cord from the heating mantle MUST be plugged into the Powermite outlet.
10. Before proceeding with the distillation have your instructor check your completed apparatus.
11. Turn on the water to the West condenser. A gentle flow is sufficient. An excessive water flow rate can disconnect a hose and cause a flood.
12. To start the distillation, turn the knob on the Powermite to a setting of 35-40. An indicator light on the Powermite unit should turn on.

The boiling point of the distillate is read from the thermometer when liquid drops from the bulb. The distillation must be stopped when only a few mL of solution remain in the distilling flask. NEVER distill a liquid to dryness. After the distillation is completed, turn off the Powermite, lower the heating mantle completely away from the distilling flask, and allow the distilling flask, distillation column, and connecting adapter to cool to room temperature.

REFLUXING

Refluxing a reaction mixture means heating it at the boiling point of the solvent. Chemical reactions occur faster at higher temperatures. The boiling point of a solution will not change significantly as long as the concentration of the solution remains constant. So heating a reaction mixture at the boiling point of the solvent, refluxing, ensures that the temperature at which the reaction occurs remains essentially constant. Loss of solvent or other volatile components is prevented by condensing the vapor that is produced and returning it to the reaction vessel. The equipment required for refluxing a reaction (see Figure labeled "Refluxing") includes a reaction flask (usually a round bottomed flask having a volume two times as large as the reaction mixture) and a condenser (West condenser). A heating mantle sized to fit the reaction flask is used as the heat source. Assemble the apparatus for refluxing as follows:

1. Clamp the reaction flask high enough above the bench top (at least 6") so that the heating mantle and its supporting ring can be lowered completely below the distilling flask. Use the vertical support rod nearest to the Powermite electrical connector.
2. Add one or two boiling chips to the reaction flask and then add the reactants using a funnel to add liquid reactants and the solvent.
3. Insert the West condenser into the reaction flask making sure that the ground glass joint is properly greased. Use a clamp to support the West condenser.
4. Connect rubber hoses (2) to the West condenser. Be sure that the hose from the water source (bottom of an aspirator) is connected to the lower connector on the West condenser. The hose from the upper connector on the West condenser should lead to the drain.
5. Use a ring clamp to support a heating mantle that exactly fits the reaction flask. The power cord from the heating mantle MUST be plugged into the Powermite outlet.
6. Before proceeding have your instructor check your completed apparatus.
7. Turn on the water to the West condenser. A gentle flow is sufficient. An excessive water flow rate can disconnect a hose and cause a flood.
8. To start the refluxing process, turn the knob on the Powermite to a setting of 35-40. An indicator light on the Powermite unit should turn on.

Once the refluxing process starts, vapor should condense in the lower 2 to 3 inches of the West condenser. A steady dripping of condensate back into the reaction flask should occur. If the hot vapor extends beyond 3 inches or so up the West condenser, the Powermite setting is too high. After the refluxing is completed, turn off the Powermite and lower the heating mantle completely away from the reaction flask, and allow the reaction flask to cool to room temperature.

Refluxing

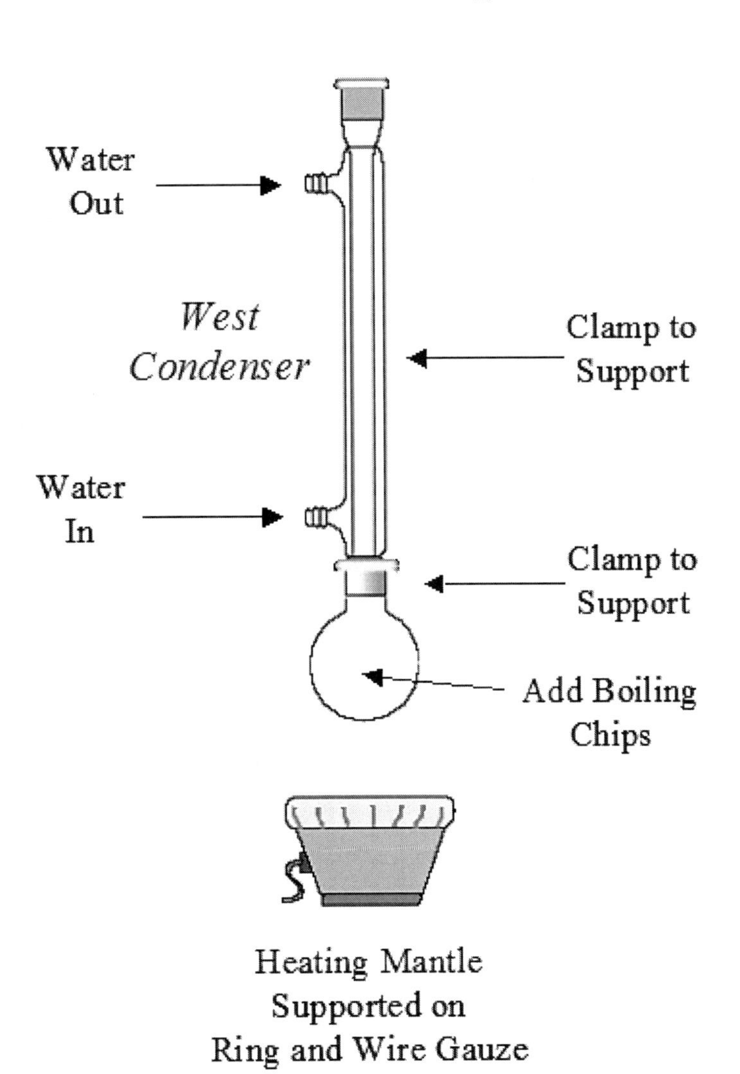

EXTRACTION

Extraction is one of the techniques that is used to separate mixtures of compounds. While there are a number of different types of extractions that are routinely carried out in a research laboratory, the type that you will carry out in this course will be a liquid-liquid extraction, hence this is the only one that we will discuss here.

Organic reactions often result in the formation of mixtures of "desired" and "undesired" products along with unreacted starting materials. Further, since most organic reactions are carried out in the presence of a solvent (A), the chemist ends up with a solution from which the desired compound(s) must be separated. Quite often, this is accomplished by extracting the reaction-mixture with another solvent (B) that is insoluble in (immiscible with) the original solvent. This technique is based on the fact that a compound (C) will distribute (partition) itself between two immiscible solvents based on how soluble it is in each of the two solvents. Since this is an equilibrium process, at a given temperature, we can write an expression for the equilibrium constant (K_p, called the distribution or partition coefficient), as shown in equation (1).

$$K_p = (g_{c,b}/v_b)/(g_{c,a}/v_a) \qquad (1)$$

The terms $g_{c,a}$ and $g_{c,b}$ represent the weights in grams of compound C dissolved in solvents A and B, respectively, while v_a and v_B represent the volumes in milliliters of solvents A and B, respectively. Thus, $g_{c,x}/v_x$ is the solubility of compound C in solvent X.

In order to achieve a good separation of the compound (C) to be removed from solvent A, the partition coefficient must have a value greater than one ($K_p > 1$). Therefore, solvent B (called the extracting solvent) must be chosen so that the solubility of compound (C) in it is greater than its solubility in solvent A. The larger K_p is, the smaller the number of extractions that will be required in order to separate compound C from solvent A. It should also be pointed out that, a larger amount of compound C will be obtained by carrying out several smaller extraction using a given total volume of the extracting solvent than by carrying out one large extraction, using the same total volume of the extracting solvent.

Liquid-liquid extractions are carried out in separatory funnels. Since the two solvents are immiscible, they will appear as two distinct layers or phases. The place at which the surfaces of the two solvents meet is called a phase boundary. The relative positions of the two solvents, i.e., which is the top or bottom layer, depend primarily on the relative densities of the two solvents. Thus, the less dense solvent will be the top layer and vise versa. However, because an extraction involves solutions rather than pure solvents, their relative positions may sometimes be interchanged. Whether or not this happens will depend on, both, the amounts and densities of the components dissolved in each layer.

The extracting solvent is usually a low boiling solvent, like diethyl ether or dichloromethane (methylene chloride). This choice is made so that the extracting solvent can be easily removed (by either a simple distillation or evaporation) from the separated compound(s) once the extraction is over.

The examples below illustrate how, both, the number of extractions and the value of K_p affect the amount of compound obtained during an extraction. For this purpose, let us suppose that we have a solution which contains 10 g of compound C dissolved in 100 mL of H_2O that we plan to extract with 100 mL of ether. Let us further assume that the solubility of compound C in water is 1.0 g/mL and that its solubility in ether is 3.0 g/mL, thus, $K_p = (3.0 \text{ g/mL})/(1.0 \text{ g/mL}) = 3$.

A Single Extraction

1. How many grams of C will be obtained if we carry out one extraction with the full 100 mL of ether?

Solution: Let x = grams of C that dissolves in the ether, therefore 10-x = grams of C that remains dissolved in the H_2O layer. Substituting into equation 1 gives equation 2. Simplifying equation 2 we get equation 3,

$$K_p = (x/100)/[(10-x)/100] = 3 \quad (2)$$

which leads to equations 4 and 5.

$$x/(10-x) = 3 \quad (3)$$
$$4x = 30 \quad (4)$$
$$x = 7.5 \text{ g} \quad (5)$$

Multiple Extractions

2. How many grams of C will be obtained if we carry out two extractions using two 50 mL portions of ether?

Solution: For the first 50 mL extraction,

$$K_p = (x/50)/[(10-x)/100] = 3 \quad (6)$$
$$2x/(10-x) = 3 \quad (7)$$
$$5x = 30 \quad (8)$$
$$x = 6.0 \text{ g} \quad (9)$$

Therefore, there are 10.0 g - 6.0 g = 4.0 g of compound C left in the water layer.

For the second 50 mL extraction,

$$K_p = (x/50)/[(4-x)/100] = 3 \quad (10)$$
$$2x/(4-x) = 3 \quad (11)$$
$$5x = 12 \quad (12)$$
$$x = 2.4 \text{ g} \quad (13)$$

Thus, the total amount of C obtained in this case is 6.0 g + 2.4 g = 8.4 g, which is larger than the 7.5 g obtained in example 1.

$K_p > 3$ and A Single Extraction

3. How many grams of C will be obtained if $K_p = 4$ and we carry out only one extraction with the full 100 mL of ether?

Solution: Proceeding as in example 1,

$$K_p = (x/100)/[(10-x)/100] = 4 \quad (15)$$
$$x/(10-x) = 4 \quad (16)$$
$$5x = 40 \quad (17)$$
$$x = 8.0 \text{ g} \quad (18)$$

Note that this amount is larger than the 7.5 g obtained in example 1.

USE OF A SEPARATORY FUNNEL

Because of its pear shape, when in use, a separatory funnel must be supported in a ring of the proper size that is clamped to ring stand at a height that is comfortable for the person using it. Use of a ring allows the funnel to be quickly inserted and removed as necessary (see Figure labeled "Extraction").

Before using the funnel, make sure that the stopcock is firmly seated, rotates freely and does not leak. Also make sure that the stopper (glass or plastic) fits tightly and does not leak. Put water in the funnel in order to test for leaks. Before you pour any liquid into the funnel, make sure that the stopcock is closed.

Since the funnel will be shaken in order to mix the two liquid layers, there must be enough room in the funnel for this mixing to occur. Thus, the funnel must never be filled completely with liquids. It is usually filled to between two-thirds and three-quarters of its capacity.

Extraction

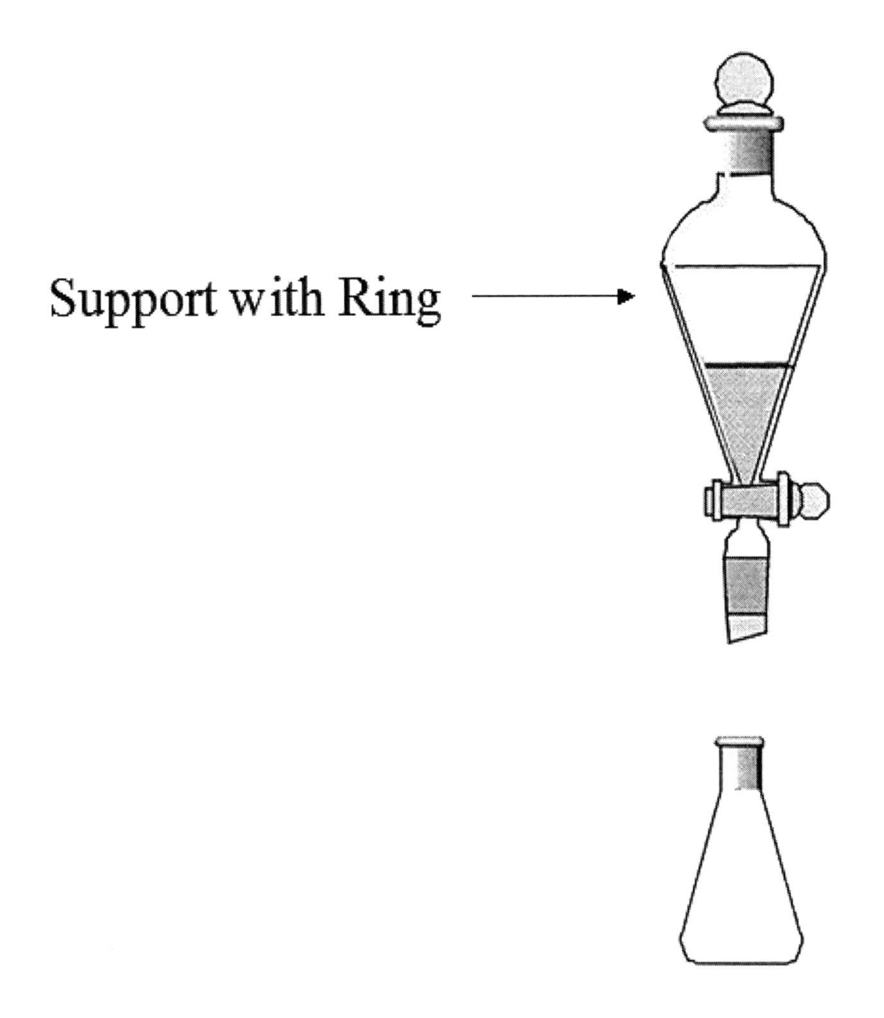

When carrying out an extraction, remove the funnel from the ring by gripping it around the neck in such a manner that the stopper is held firmly in place with the heel of one hand and the stopcock is held by the other hand (see photograph). Invert the funnel and immediately open the stopcock in order to release any pressure which might have built up. *During this venting process always make sure that the tip of the funnel is pointing up and away from everyone* . Close the stopcock and with the funnel held at an angle, as shown in the photograph, shake it vigorously for about 15 to 30 seconds. Slowly

open the stopcock in order to release any pressure as indicated above. Repeat the shaking and venting steps several more times. Turn the funnel up right, replace it in the ring and immediately loosen and then remove the stopper. Allow the layers to separate completely and then open the stopcock so that the bottom layer drains off into a receiver. The top layer is removed from the funnel by pouring it out through the mouth of the funnel into another receiver.

VACUUM FILTRATION

Vacuum filtration is used to collect or isolate solids (crystals or precipitates) from a solution. A vacuum is used to speed the filtering process. The apparatus consists of a filter flask (250 mL side-arm flask), a Buchner funnel inserted through a one-holed rubber stopper, and a piece of filter paper (see the Figure labeled Vacuum Filtration). Clamp the 250 mL side arm flask to a vertical rod. Connect the side-arm of the flask to the nearest aspirator using a length of rubber tubing. Insert the rubber stopper holding the Buchner funnel into the top of the side arm flask. Obtain a piece of filter paper of the appropriate size from the stockroom, and place the filter paper in the Buchner funnel. The paper should fit in the bottom of the Buchner funnel without folding. Just before you are ready to start the vacuum filtration, turn on the water at the aspirator to produce a vacuum. Carefully pour a few mL of the solvent or liquid you are using into the Buchner funnel to completely wet the filter paper in the funnel. The vacuum should seal the filter paper down to the flat surface of the Buchner funnel. Gently swirl the flask containing the solution to be filtered in order to suspend the solids in the solution. Then carefully pour the solution containing suspended solids into the Buchner funnel. Rinse out the flask with a few mL of solvent and pour the rinse solution into the Buchner funnel too. The solid can be dried by keeping the aspirator running for a few minutes. After the solid has air dried, you can carefully remove the filter paper and the solid from the Buchner funnel using your spatula. The solid can be completely dried by placing the filter paper and solid in a plastic weighing boat, and then storing the weighing boat in your drawer until the next laboratory period.

Vacuum Filtration

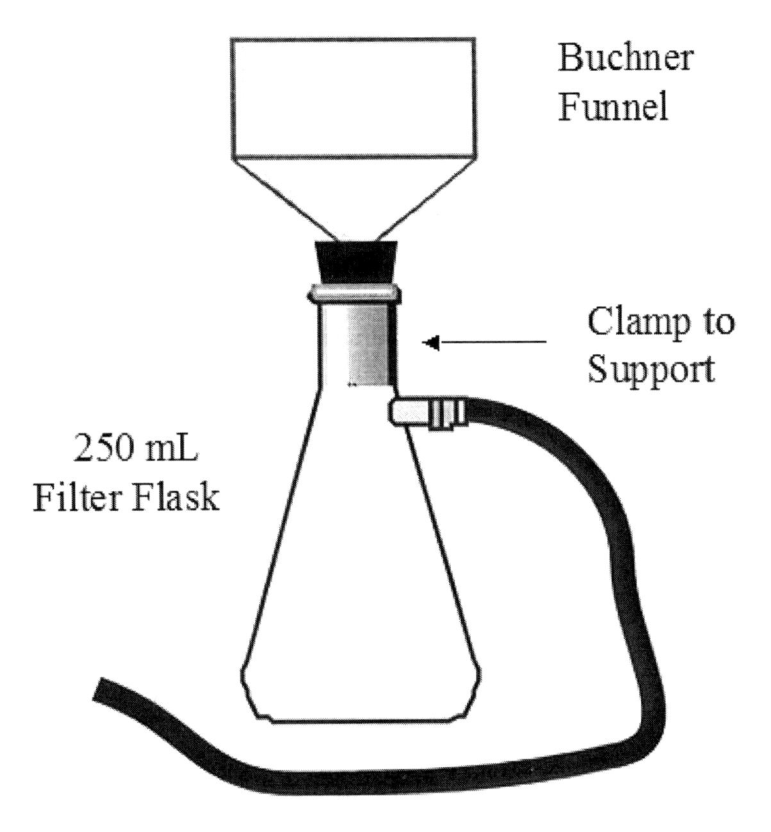

RECRYSTALLIZATION

The most common method for the purification of solids is called recrystallization. Solids are obtained from natural sources or reactions and often contain small amounts of other substances (impurities). In order to remove the impurities, the impure solid is dissolved in a hot solvent, and the solvent is then allowed to cool. The pure solid forms crystals as the solvent cools, but the impurities remain dissolved. Finally, filtration (usually suction filtration with a Buchner funnel) is used to separate the crystals from the solvent. The process is shown in diagrams below x represents the solid which will be obtained in the pure state, and o is the impurity.

1. Place solid in an Erlenmeyer flask (or beaker)

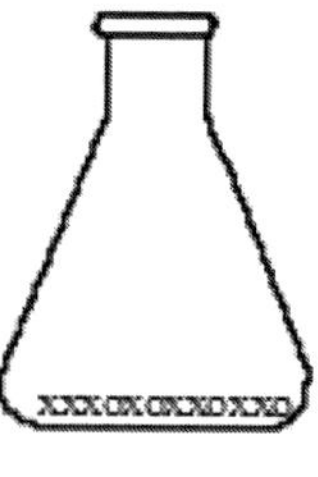

2. Add solvent

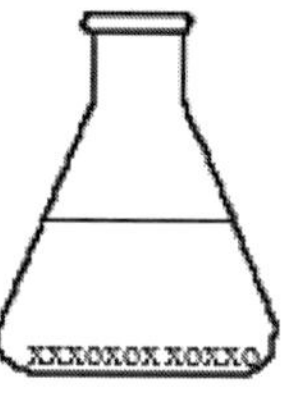

3. Heat the solution to dissolve the solids.

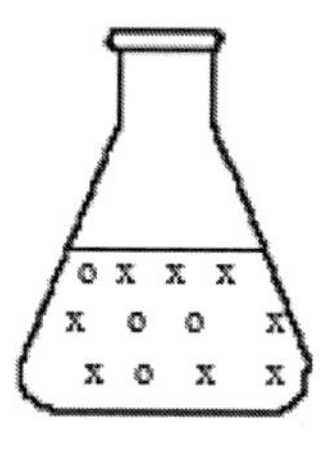

4. Allow the solution to cool. Impurity remains in solution

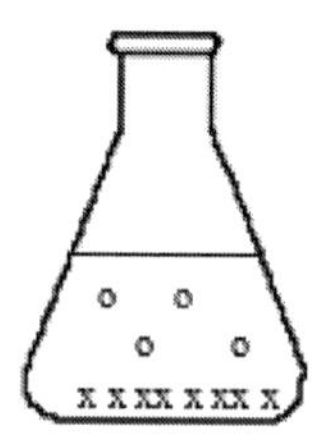

Because the concentration of the impurity o is small, and the solution is saturated with respect to x, cooling causes x to crystallize and leaves o dissolved in the solvent.

Although the procedure is outlined above, there are several practical problems. The chief problem is the selection of the recrystallization solvent. Ideally, the solvent should dissolve any quantity of x when it is hot, and should not dissolve any x when it is cold. This is impossible, and one consequence is that some of the x remains in solution after cooling, that is, purification is accompanied by a loss of material. Also, the solvent should dissolve o when it is hot or cold, so that o does not precipitate with x. The choice of solvent may take considerable effort with trials of several solvents. Sometimes, mixtures of solvents (ethanol/water, methanol/water) work better than pure liquids. Generally, low boiling solvents such as ether or dichloromethane are not useful for recrystallization. This is because the solubility of the solid in the cold solvent and boiling solvent does not vary enough, because the temperature range is too small. On the other hand, very high boiling solvents are avoided because they are hard to remove from the solid after filtration.

Sometimes the impurity o is insoluble in the hot solvent. In that case, the solution is filtered quickly, using gravity filtration, while it is hot (special heated funnel holders are used), and after insoluble o is removed, the solution is allowed to cool.

GAS CHROMATOGRAPHY

Mixtures of organic compounds can be separated by the differential partitioning of these compounds between a stationary phase and a mobile phase. The various types of chromatography used by organic chemists such as column chromatography, thin layer chromatography (TLC), and gas chromatography (GC) are distinguished from each other by the nature of the two phases that are used. In gas chromatography the mobile phase is a gas (usually helium) and the stationary phase is a very high boiling liquid (polydimethysiloxane, HP-1) coated on the inside of a long (5 m) glass capillary tube.

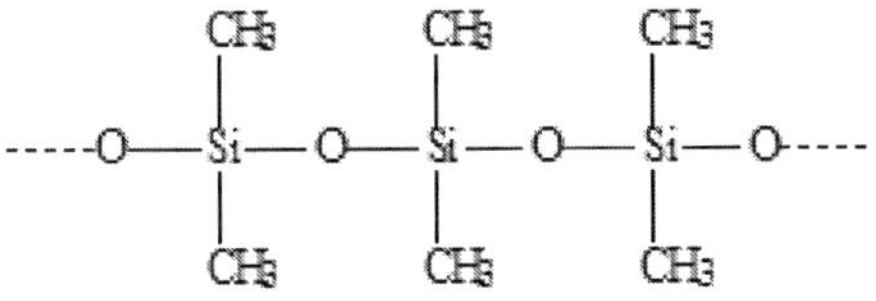

```
        CH3        CH3        CH3
         |          |          |
-----O---Si---O----Si---O----Si---O-----
         |          |          |
        CH3        CH3        CH3
```

polydimethylsiloxane

Using GC a very small volume (1 μL) of a dilute solution of a mixture of organic compounds can be completely separated because each compound in the mixture has a different affinity for the stationary phase and thus will take a different time, called the retention time (RT), to pass through the column. The sample mixture is initially injected using a microliter syringe into the injection block of the gas chromatograph where the sample is vaporized at high temperature and then swept into the capillary column by a stream of the carrier gas (He), the mobile phase. The column is located in an oven so the temperature of the column can be controlled or systematically increased during the run. Material emerging from the column, the effluent, is detected by a device called a flame ionization detector. In the flame ionization detector the effluent is burned in a hydrogen flame, producing ions which are counted electronically. A graph of the number of ions produced versus elapsed time (since injection of the sample) is generated by the recorder or integrator attached to the gas chromatograph. A schematic diagram of a gas chromatograph is shown below.

Each compound will have a characteristic retention time for a particular column, oven temperature profile, and carrier gas flow rate (instrument parameters). Thus the

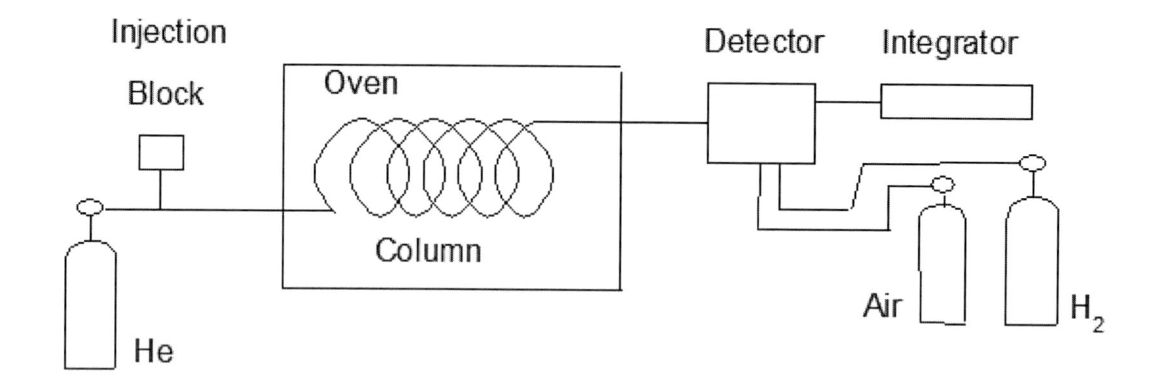

identity of a specific peak in a gas chromatogram can be assigned by comparing and correlating the retention time of the peak with the retention times of known compounds that were analyzed on the same instrument under identical conditions (instrument parameters). The relative amount of a particular compound present in a mixture is related to the integrated area of the peak corresponding to that particular compound in the gas chromatogram of the mixture. The Agilent 7890A Gas Chromatography system prints out a table containing not only the retention time (RT) of each peak detected but also the integrated area (AREA) calculated for each peak. Usually the peak with the largest area and shortest retention time is associated with the solvent, dichloromethane (CH_2Cl_2), used in many of the GC analyses you perform. In calculating the relative amounts of each compound present in your original samples, you must neglect or exclude from your calculations any material due to the solvent, CH_2Cl_2, that you employed in making up the samples you actually used for the GC analysis.

Examples of gas chromatograms and their analysis can be found in the "Sample Notebook Write-Up" section of the Introduction.

THIN LAYER CHROMATOGRAPHY

Thin layer chromatography is a form of chromatography in which the stationary phase is a solid adsorbent (e.g. silica gel ($SiO_2 \bullet xH_2O$) or alumina ($Al_2O_3 \bullet xH_2O$)) supported on a thin sheet of glass, metal, or plastic. The mobile phase is usually an appropriate solvent contained in a closed bottle (developing chamber). TLC is frequently used in the laboratory by chemists to determine roughly the number components in a mixture. It is also used to monitor the progress of a chemical reaction or purification by column chromatography. TLC can also serve as a simple and quick way to determine the identity of a compound, when an appropriate reference sample is available. The apparatus for TLC is illustrated in Figure 1.

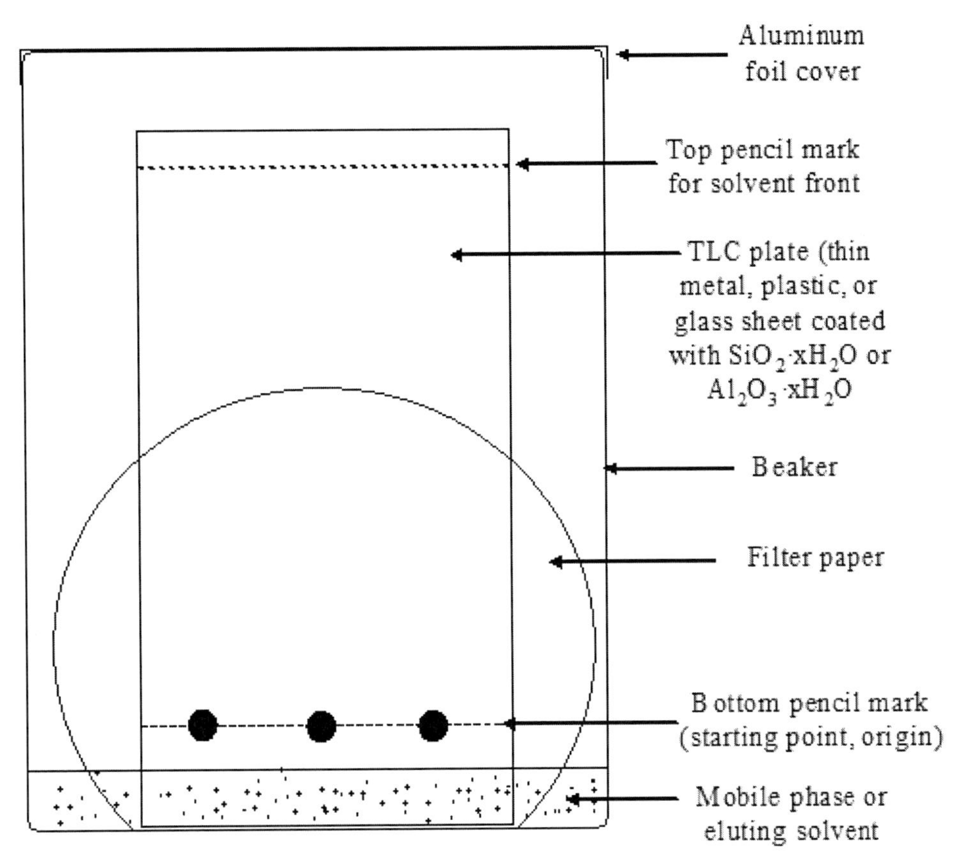

Figure 1

Running a TLC involves dissolving the sample to be analyzed in a volatile solvent and spotting this sample solution on a TLC plate (Figure 1). A known reference sample e.g. starting material, or a previously synthesized product can be spotted side by side on the same plate for easy comparison. To ensure that the solutions are spotted at approximately the same distance from the bottom of the plate, a pencil mark is made on the plate about 5-6 mm from the bottom of the plate (do NOT use a pen). In addition, a pencil mark is made about 3 mm from the top of the plate to assist in determining when the plate should be removed from the developing chamber and locating the position of solvent front on the developed plate . Once the plate has been spotted, it is placed in the developing chamber, which contains a shallow level of the mobile phase or eluting solvent. It is essential that the spotted level on the TLC plate is above the solvent level in the developing chamber. The solvent or liquid mobile phase then travels up the plate by capillary action taking components of the mixture with it. Because components of the mixture have different affinities for the stationary phase versus the mobile phase, they are separated as the solvent moves them up the plate. The more polar components that interact most strongly with the polar silica gel or alumina stationary phase are carried more slowly by the solvent. On the other hand, less polar or non-polar components interact more weakly with the polar stationary phase, so they spend less time on the stationary phase and more time in the mobile phase. Thus they are carried further and so will end up closer to the top of the plate. It is important to note that if the eluting solvent or mobile phase is made more polar, it can carry the components further. Hence, choosing an appropriate solvent for a particular TLC analysis is very important. Once in the developing chamber, the plate is allowed to develop until the solvent front is almost at the top of the plate (~ 3 mm from the top). At this point, the plate is removed from the developing chamber and the solvent allowed to evaporate from the plate. If a pencil mark has been marked close to the top of the plate, the plate can be removed as soon as the solvent front just reaches this mark.

The developed plate can be visualized by various means to determine the position of each component of the mixture on the plate. Two commonly used techniques are visualization under an ultra violet lamp and visualization in an iodine chamber. In the later case, the iodine vapor in the chamber can oxidize some organic compounds or can be momentarily adsorbed on them so that they appear brown. The spots that appear should be outlined with a pencil immediately upon removing the plate from the iodine chamber, as the iodine spots gradually fade. In the case of visualization under UV, fluorescent materials can appear as bright spots against a dark stationary phase background. On the other hand, if the plate is pre-coated with a fluorescent material, when it is viewed under the UV lamp, the organic components quench or block the fluorescence in the spots they occupy. Hence, the area covered by each component or spot on the plate appears as a dark spot against a bright fluorescent background. The dark area can then be outlined with a pencil while visualizing under the UV lamp. [**Warning: Do Not expose Eyes and Skin to short wave ultraviolet light. UV light can be harmful to unprotected eyes and skin !!!**]

How far up the plate a compound travels is represented by the R_f (retention factor, also sometimes called the "ratio to front"). The R_f value for a particular compound is affected by a number of factors including the type of stationary phase and mobile phase, stationary phase thickness, amount of analyte spotted, temperature, and the properties of the compound itself. The R_f value of a component of a mixture can be compared to that of an authentic sample run on the same plate.

In general more polar solvents cause compounds to migrate further up a TLC plate. This propensity is referred to as the eluting power of the solvent. The order of eluting power for solvents that can be used in TLC is: water > methanol > acetonitrile > ethanol > acetone > isopropyl alcohol > ethyl acetate > chloroform > diethyl ether >

dichloromethane > toluene > alkanes (hexane, ligroin, petroleum ether). Mixtures of two solvents are sometimes used in order to fine tune the eluting power of the solvent system. Polar organic compounds interact more strongly with the polar silica gel or alumina stationary phases than do non-polar organic compounds. So non-polar organic compounds will tend to migrate further than polar organic compounds in TLC. The relative polarities of some types of organic compounds are: carboxylic acids > alcohols > amines > aldehydes or ketones > esters > ethers > halocarbons > aromatic hydrocarbons > conjugated dienes > alkenes > alkanes.

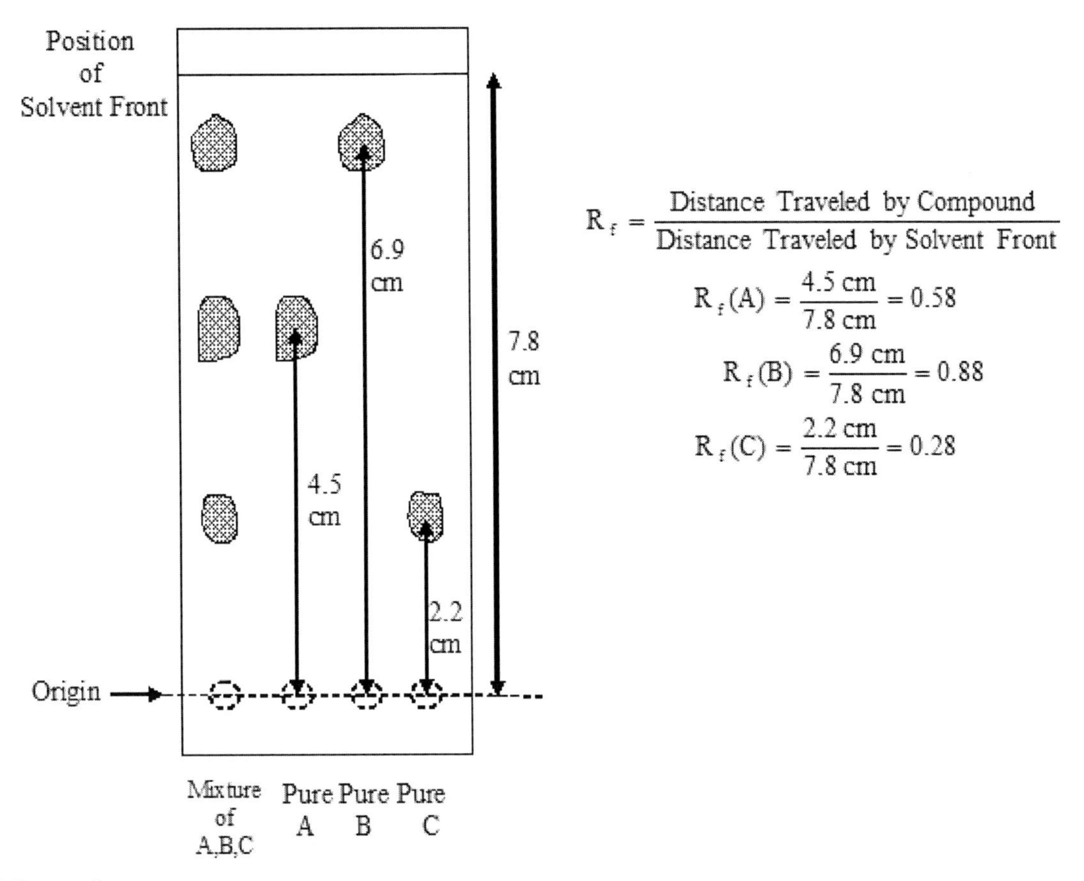

Figure 2

The R_f value of a compound is calculated as the distance traveled by the compound divided by the distance traveled by solvent front, as shown in Figure 2. The center of each spot on the plate is used in calculating its R_f value.

In this course, we will use a silica gel stationary phase containing a fluorescent indicator pre-coated on a thin sheet of aluminum or plastic. This sheet can easily be cut to the required size with a pair scissors.

Sample preparation: Fill the narrow end of a Pasteur pipette to a depth of about 1.0 cm with your solid or liquid sample. Dissolve this solid or liquid in 8-10 drops of dichloromethane in a small test tube. Apply one small spot of this sample solution to your TLC plate using a fine glass capillary or plastic pipette tip by touching the end of the capillary or pipette tip to the plate about 6 mm from the bottom of the plate. Two to four different samples can be analyzed on a single TLC plate.

INFRARED SPECTROSCOPY

The absorption of electromagnetic radiation in the infrared region can give information about the types of bonds in organic molecules. Thus infrared spectroscopy can be extremely useful in confirming the identity of an organic molecule or in determining functional groups present in an unknown organic molecule.

The energy of a photon of electromagnetic radiation is related to its wavelength and frequency as follows

$$E = h \cdot \nu = h \cdot \frac{c}{\lambda}$$

where h is Plank's constant (6.626×10^{-34} J·sec), ν is the frequency (units of cycles per second or Hz) and c is the speed of light (3.00×10^{10} cm/sec) and λ is the wavelength (units of cm). Most infrared spectra of organic molecules are recorded not in units of wavelength or frequency but in a derived unit called the wave number, $\bar{\nu}$ (units of cm^{-1}) where

$$\bar{\nu} = \frac{1}{\lambda} = \frac{\nu}{c}$$

Note: wavelength (μm) × wave number (cm^{-1}) = 10,000 (μm/cm)

The infrared region of the electromagnetic spectrum that organic chemists use is from 4000 cm^{-1} to 600 cm^{-1} with a strong emphasis on the region from 4000 cm^{-1} to 1500 cm^{-1}.

The infrared spectrum of an organic molecule is recorded as a graph of the %T (percent transmittance) of the molecule as a function of the wave number of the infrared light. The percent transmittance is defined as the fraction of the light that is absorbed by the molecule expressed as a percentage. Thus if light of a particular wavelength with intensity I_0 is shined on a sample and after the light has passed through the sample the intensity of the light is reduced to I, the %T of the sample for that particular wavelength of light is

$$\% T = 100 \times \frac{I}{I_0}$$

So if the sample absorbs none of the light at a particular wavelength, the %T for that wavelength will be 100%. At the other extreme, a sample that absorbs all the light at a particular wavelength will have a %T of 0% at that particular wavelength. So absorption is roughly inversely proportional to transmittance.

The infrared spectra of a number of different organic compounds are shown below. Each peak (region of lower %T) is called a band. Each band in the infrared spectrum of an organic compound is characterized in three ways, position, intensity, and shape. The position of a band is the wave number at which it has minimum %T. The infrared spectrometers you will use usually print out the position for most of the bands in the infrared spectrum. The intensity of a band is usually described using the words, strong, moderate (medium), or weak A strong band has a very low %T at its position, while a weak band has a large %T (not very much smaller than the baseline) at its position. As a rough guide bands that have a %T in the range of 75-100% are weak, 35-70% are moderate, and 0-30% are strong. Thus the intensities of the bands in the region of 1700-1750 cm^{-1} in the IR spectra of cyclohexanone, methyl heptanoate, and heptanoic acid (see spectra below) are all strong. The shape of a band is usually described using the words sharp, medium, or broad. A sharp band occurs over a very narrow range of wave numbers, while a broad band occurs over a much wider range of wave numbers. In the spectra shown below the band at 3331 cm^{-1} in the IR spectrum of cyclohexanol is a broad band where as the bands at 2932 cm^{-1} and 2855 cm^{-1} are sharp bands. Essentially all the bands in the IR spectrum of phenylacetonitrile (see below) are also sharp.

The position of a band in the infrared spectrum of an organic compound is related to the identities of the atoms and the type of bond holding them together. For the infrared spectroscopy of organic molecules the bond holding two atoms together can be viewed as a simple spring joining two masses together. Using simple physics, the frequency (ν) or the wave number ($\bar{\nu}$) at which such a system vibrates can be described using three parameters, M_1, M_2, and k as follows:

$$\bar{\nu} = \left(\frac{1}{2\pi c}\right)\sqrt{\frac{k}{\mathrm{M}}}$$

where M, the reduced mass is defined as

$$\mathrm{M} = \frac{M_1 \cdot M_2}{M_1 + M_2}$$

M_1 and M_2 are the masses of each of the atoms involved in the bond, k is the force constant associated with the bond (related to the bond strength) and c is the speed of light. So the position (wave number) of a C−H bond would be higher than that of a C−O bond since the mass of O is much greater than the mass of H. Similarly the position of a C≡N (triple bond) would be higher than that of a C=N (double bond) which would be higher than that of a C−N (single bond) since triple bonds are stronger (larger k) than double bonds which are in turn stronger than single bonds between the same two atoms.

The characteristic positions (wave number), intensities, and band shapes of the infrared bands of a number of common types of bonds found in organic molecules are summarized in the following table.

Bond Type	Position (cm^{-1})	Intensity / Shape
O−H	2500−3600	moderate-strong/broad
N−H	3300−3500	moderate/sharp
C−H	2700−3300	variable/sharp
C≡N	2200−2300	moderate-strong/sharp
C≡C	2100−2300	variable/sharp
C=O	1650−1850	strong/sharp
C=C	1600−1680	weak-moderate/sharp
C−O	1000−1300	strong/medium-sharp

Clearly there is considerable overlap in the band positions of single bonds (C−H, N−H, and O−H), double bonds (C=O and C=C), and triple bonds (C≡N and C≡C). Within each of these bond types there are also differences the positions of bands. Thus for C−H bonds the exact position in the above range depends on the nature of the C−H bond, in particular the hybridization of the carbon. Thus a C−H that is part of an alkyne group (C≡C−H, sp hybridization for carbon) occurs at around 3300 cm^{-1}, while a C−H that is part of an alkene or aromatic group (C=C−H, sp^2 hybridization for carbon) occurs at 3010-3100 cm^{-1}, and the C−H of an alkane (C−C−H, sp^3 hybridization for carbon) occurs at 2700-2970 cm^{-1}.

The IR spectra of cyclohexane, cyclohexene, and 1−pentyne are shown below. The only significant bands in the IR spectrum of cyclohexane above 1600 cm^{-1} are at 2928 and 2853 cm^{-1}. These band are associated with vibrations of the CH_2 groups in the molecule. The IR spectrum of cyclohexene is very similar to that of cyclohexane (in the region between 2800 and 3000 cm^{-1}), except there are additional sharp bands at 3023 and 1652 cm^{-1}. These bands are associated with the alkenyl group present in cyclohexene

(C=C−H). The strong, sharp band at 3023 cm^{-1} is attributable to the C−H of the C=C−H group, while the weak, sharp band at 1652 is attributable to the C=C of the C=C−H group that is present in cyclohexene but not cyclohexane. The IR spectrum of toluene is similar to that of cyclohexene with sharp bands at 3026 and 1605 cm^{-1} corresponding to the C-H and C=C of the aromatic ring respectively. The bands at 2920 and 2823 cm^{-1} are associated with the C-H's of the methyl group The IR spectrum of 1−pentyne is also similar to that of cyclohexane in the region of 2800-3000 cm^{-1} (where the vibrations associated with the CH_3 and CH_2 groups occur) except in this case there are additional sharp bands at 3307 and 2120 cm^{-1}. These bands correspond respectively to the C−H and the C≡C of the C≡C−H group in the molecule.

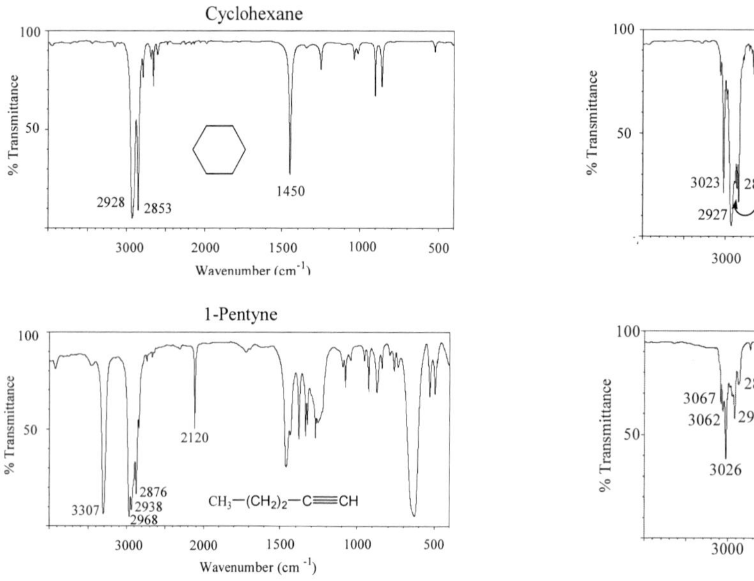

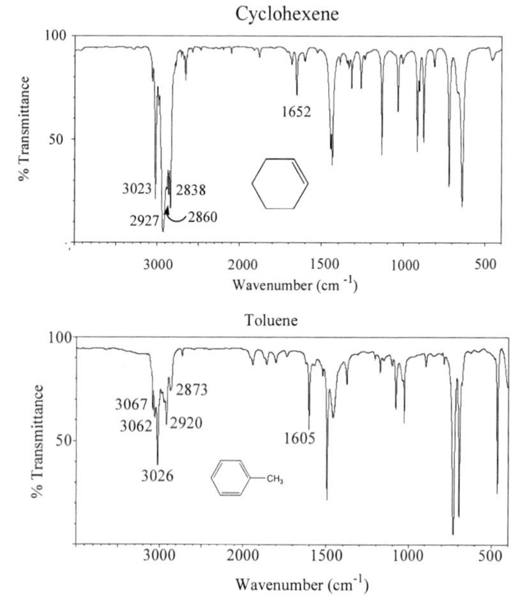

Whereas the C−H bands of alkanes, alkenes, and alkynes all are sharp, the O−H bands of alcohols, phenols, carboxylic acids, and the N−H bands of amines (primary and secondary) are all broad, as can be seen in the IR spectra of cyclohexanol, n−pentylamine, heptanoic acid, and phenol shown below. For cyclohexanol and phenol there are broad, strong O−H bands at 3331 and 3229 cm^{-1} respectively, while in n-pentylamine (a primary amine) there are two weak, broad N−H bands at 3370 and 3292 cm^{-1}. Secondary amines generally have only one band in this region. The IR spectra of cyclohexanol and n−pentylamine also have strong, sharp bands in the region of 2800 to 3000 cm^{-1} attributable to the CH_3 and CH_2 groups in these molecules. The O−H band in carboxylic acids (see heptanoic acid below) tend to be even broader than the O-H bands in alcohols and phenols, frequently extending over the entire region from 2400 to 3600 cm^{-1}. Thus the IR spectrum of heptanoic acid has the sharp bands of the CH_3 and CH_2 groups (2969, 2933, and 2861 cm^{-1} superimposed over a very broad O–H band. This effect can be more easily seen by comparing the IR spectra of heptanoic acid with that of methy heptanoate, which as an ester does not contain an O–H group but does have the bands associated with CH_3 and CH_2 groups at 2957, 2934, and 2861 cm^{-1}.

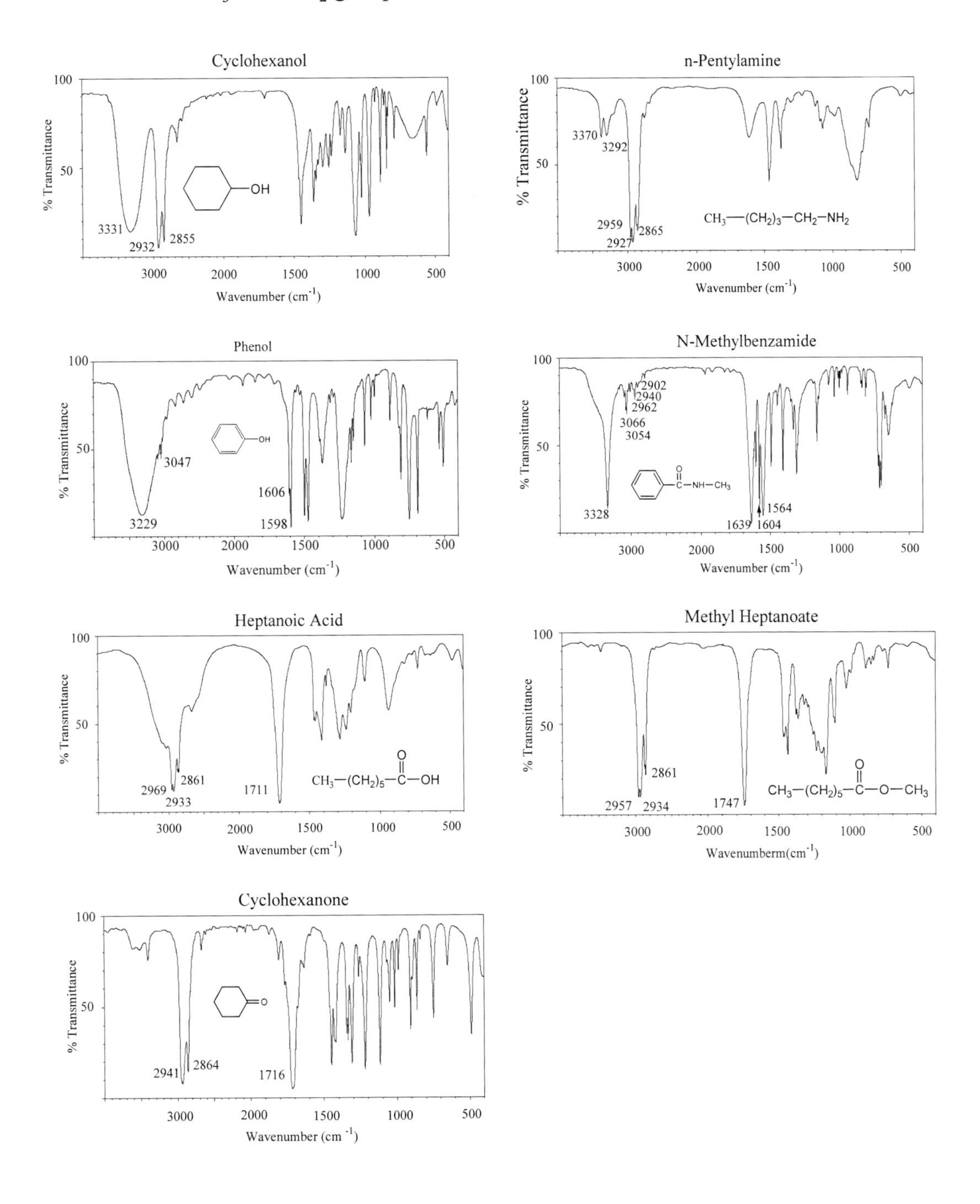

The carbonyl groups (C=O) of aldehydes, ketones, carboxylic acids, esters, and amides all have strong, relatively sharp bands in the 1650 to 1800 cm^{-1} region. This band is found at 1716 cm^{-1} in cyclohexanone (see above), 1711 cm^{-1} in heptanoic acid (see above), 1747 cm^{-1} in methyl heptanoate (see above), and 1639 cm^{-1} in N-methylbenzamide (in amides this band is called the "amide I" band). The exact band position of the carbonyl (C=O) band depends on the nature of the carbonyl group as indicated in the following table.

Carbonyl Compound	C=O Band Position (cm^{-1})
Ester	1730-1750
Aldehyde	1720-1740
Ketone	1700-1730
Carboxylic Acid	1700-1720
Amide	1650-1690

So esters generally have their C=O band at higher wave numbers than do aldehydes, which have their band at higher wave numbers than ketones, which have their band at higher wave numbers than acids, which have their band at higher wave numbers than amides. However, there clearly is some overlap in these ranges. The position of the carbonyl band also depends on whether it is conjugated to an aromatic ring or alkene. Conjugation tends to lower the position of the carbonyl band.

The IR spectrum of N-methylbenzamide also shows a strong band at 3328 cm^{-1} attributable to the amido N-H, and weak bands at 3066 and 3054 cm^{-1}, as well as at 2962, 2940, and 2902 attributable to the aromatic and aliphatic C-H's in the molecule respectively. The sharp band at 1604 cm^{-1} is attributable to the aromatic C=C's in the molecule. Amides that have a hydrogen attached to the nitrogen also show a strong broad band, called the "amide II" band that is generally found in the region 1550 to 1650 cm^{-1} (it is always located at a lower wave number than the carbonyl band. The strong, broad band at 1564 cm^{-1} is the "amide II" band.

Nitriles, compounds containing a C≡N bond, all have a characteristic sharp, moderate/strong band in the 2200 to 2300 cm^{-1} region, the same region where alkynes (C≡C) show sharp but weak/moderate bands. The moderate, sharp band at 2252 cm^{-1} in the IR spectrum of phenylacetonitrile (see below) is attributable to the nitrile group in this compound. Sharp aromatic C-H bands at 3091, 3067, and 3035 cm^{-1}, aliphatic C-H (from the CH_2 group) at 2953 and 2921 cm^{-1}, as well as an aromatic C=C band at 1603

cm^{-1} are also present.

Nitro (NO_2) groups have two strong, moderately sharp bands, one between 1500 and 1600 cm^{-1} and the other between 1300 and 1400 cm^{-1}. These bands occur at 1527

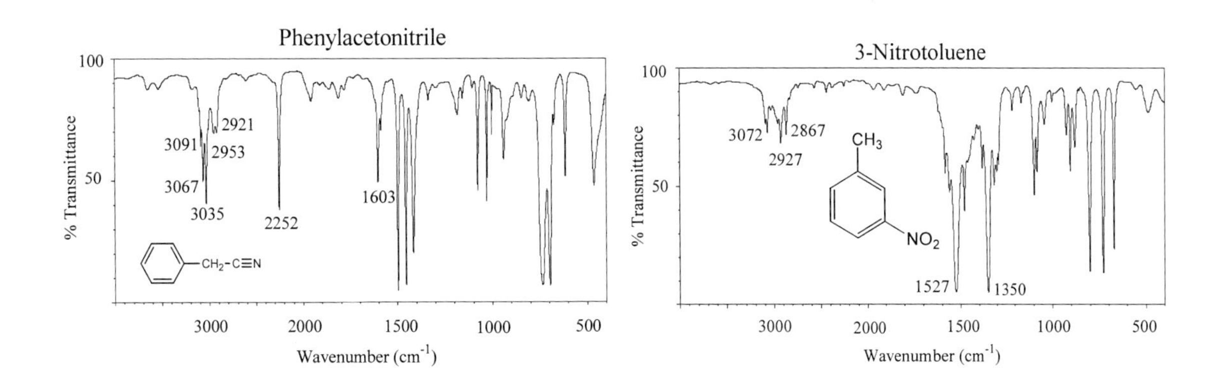

and 1350 cm^{-1} in the IR spectrum of 3-nitrotoluene (see above). A weak, sharp band attributable to the aromatic C-H at 3072 cm^{-1} and weak sharp bands attributable to the CH_3 group at 2927 and 2867 cm^{-1} are also present. The aromatic C=C band is not visible.

All IR spectra were reproduced from the Integrated Spectral Data Base System for Organic Compounds (SDBS web: http://riodb01.ibase.aist.go.jp/sdbs/cgi-bin/cre_index.cgi?lang=eng). This searchable data base contains IR, NMR, and MS spectra for a large number of organic compounds.

Additional examples of IR spectra and their interpretation can be found in the "IR Tutor" software on the computers in the Undergraduate Chemistry Computer Laboratory. IR Tutor includes IR spectra for the following compounds: n-hexane, 1-hexene, 1-heptyne, 2,3-dimethylbutane, toluene, 1-cyanoheptane, 1-hexanol, 1-aminohexane, heptaldehyde, 3-heptanone, heptanoic acid, ethyl acetate, and butyric anhydride.

PROTON NUCLEAR MAGNETIC RESONANCE SPECTROSCOPY

Nuclear Magnetic Resonance (NMR) spectroscopy is a powerful tool that is routinely used in the determination of the structure of organic compounds. The most commonly used type of NMR spectroscopy is proton NMR spectroscopy (^{1}H NMR), and that is the only kind we will discuss here. In discussing NMR spectroscopy the hydrogen atoms in a molecule are often referred to as protons.

In NMR spectroscopy, the sample is placed in a strong magnetic field and electromagnetic radiation in the radiofrequency range is applied. Each proton has a magnetic moment and will have two observable energy states (sometimes called spin states) in a magnetic field. The magnetic moment associated with each spin state is either aligned with (↑) or opposite to (↓) the magnetic field. The energy separation between the two spin states is a function of the magnetic field strength as shown in Figure 1.

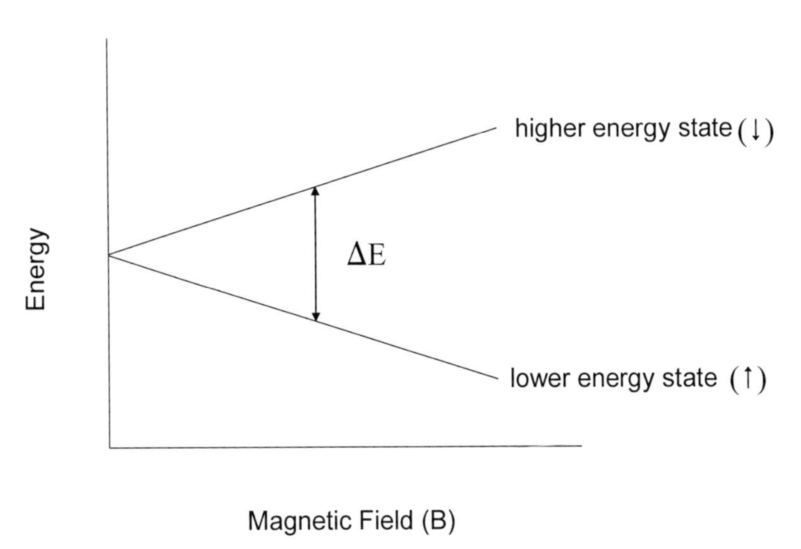

Figure 1

Absorption of energy (or more generally, transitions between these spin states) can occur when the frequency of the applied electromagnetic radiation corresponds to the energy separation (ΔE) between the spin states. This relationship can be expressed by the equation

$$\Delta E = h\nu = h\gamma B/2\pi$$

where h is Plank's constant, γ is the magnetogyric ratio (a constant characteristic of the particular nucleus (in this case a proton)), and B is the magnetic field strength.

In a constant magnetic field, each different type of proton in a molecule will generally absorb electromagnetic radiation at a slightly different frequency. The exact absorption frequency is proportional to the total magnetic field at that proton. The total magnetic field at any proton is the sum of the applied magnetic field (from the NMR spectrometer) and any local magnetic fields. The local magnetic fields are for the most part generated by the electrons in the molecule.

In modern NMR spectrometers, the radiofrequency energy is delivered as a short "pulse" having a bandwidth that includes all the frequencies necessary for the absorption of energy by all the protons in the sample. The resulting signal is transformed by a mathematical process called a Fourier transform to give the NMR spectrum in the form we normally see it.

Examples of ^{1}H NMR spectra are shown below (Figures 2-11). In each spectrum the vertical axis represents the intensity of the signals, and the horizontal axis represents the "chemical shift " (δ, in units of ppm). The chemical shifts are directly related to the frequencies required for absorption as compared to a particular standard frequency. The horizontal axis is sometimes considered to be related to the effective magnetic field felt by the protons. All the spectra shown below are computer generated simulations at 300 MHz showing the major couplings.

A ^{1}H NMR spectrum of an organic compound provides the following information:

- Chemical Shift. The positions of the signals (peaks) indicate what types of protons (hydrogen atoms) are present in the sample.

- Integration. The relative areas under the signals (peaks) indicate the relative number of protons (hydrogen atoms) of each kind that are present in the sample.

- Spin-Spin Splitting. The presence (or absence) of splitting of the signals, and the nature of the splitting often indicates how may protons (hydrogen atoms) are present on neighboring carbon atoms.

Chemical Shift: The chemical shift (δ, in units of ppm) is the position on the horizontal axis. The reference standard for ^{1}H NMR is tetramethylsilane (TMS, $Si(CH_3)_4$) which has twelve equivalent protons. The protons of TMS are assigned a chemical shift of 0 (δ = 0, 0 ppm), and the positions of protons in other compounds are defined by how far they are "shifted" from those of TMS. The protons in organic compounds generally have signals having chemical shifts in the range δ = 0 to 12. The numerical value of the chemical shift of a peak is calculated as shown in equation 1.

$$\delta = \text{chemical shift (in ppm)} = \frac{\text{number of Hz between TMS and peak in question}}{\text{operating frequency of the spectrometer (in MHz)}} \quad (1)$$

The δ scale defined in this way is convenient because it gives the same chemical shift values for the same compounds with any spectrometer. In contrast the actual frequencies depend on the strength of the magnetic field, and spectrometers of different magnetic field strength are available.

For historical reasons, δ values increase from right to left on the NMR chart. The relative positions of NMR signals or peaks are often described as being "upfield" or "downfield" from one another. "Upfield " means shifted to the right (lower δ values), and "downfield" means shifted to the left (higher δ values).

The chemical shift of an NMR signal gives information about the chemical environment of the protons responsible for the signal (i.e., aromatic as opposed to aliphatic protons).

The ^1H NMR spectrum of isopropylbenzene, shown in Figure 2, contains three groups of signals or peaks. These signals correspond to the three different types of protons found in isopropylbenzene (aromatic, benzylic, and methyl protons) and occur at about δ = 7.2, δ = 2.7, and δ = 1.6 respectively.

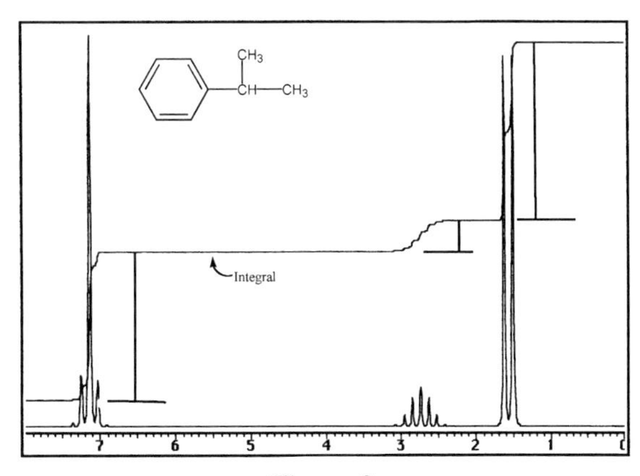

Figure 2

Table 1 shows typical values for the chemical shifts of protons in various functional groups. In general, protons attached to sp^3 hybridized carbons are usually observed at δ = 0-5, alkene protons (attached to sp^2 hybridized carbons) are usually at δ = 4.5-6.5, and aromatic protons are usually at δ = 6.0-8.5. A proton directly attached to a carbonyl group (as in an aldehyde) is usually at δ = 9-10, while a proton directly attached to a sp hybridized carbon (of an alkyne) is at δ = 2-3. The chemical shifts of protons attached to electronegative atoms (ROH, RNH_2, RCO_2H, etc.) are quite variable. NMR chemical shifts are influenced strongly by the electronic effects of nearby substituents. In general, protons attached to a carbon atom that has an electronegative atom or electron withdrawing group attached to it have a downfield chemical shift (towards higher δ values) compared to aliphatic protons.

Table 1 Chemical Shifts of Hydrogen Atoms in Various Functional Groups

Functional Group (**H** shown in Bold)	Chemical Shift δ (ppm)	Functional Group (**H** shown in Bold)	Chemical Shift δ (ppm)
TMS		Nitroalkanes	
H_3C—Si(CH_3)—CH_3 with CH_3 above and below Si	0.0	R—C(R')(NO_2)—**H**	4.2 - 4.6
Alkanes		Alkyl Chlorides	
R—CH_3	0.9	R—C(R')(Cl)—**H**	3 - 4
R—CH_2—R'	1.3	R—C(Cl)(Cl)—**H**	5.8
R—CH(R')—R"	1.5	Alkyl Bromides	
Alkenes		R—C(R')(Br)—**H**	2.4 - 4
R—C(R')=C(R")—**H** (vinyl)	4.5 - 6.5	Alkyl Iodides	
R—C(R')=C(R")—CH_3 (allyl)	1.7	R—C(R')(I)—**H**	2 - 4
Alkynes		Alcohols, Ethers, Phenols	
R—C≡C—**H**	2 - 3	R—O—**H**	1 - 5.5
R—C≡C—CH_3	1.8		
Aromatic			
Benzene ring—**H**	6 - 8.5		
Benzene ring—C(R)(**H**)—R' (benzyl)	2.2 - 3		

Functional Group (**H** shown in Bold)	Chemical Shift δ (ppm)	Functional Group (**H** shown in Bold)	Chemical Shift δ (ppm)
Alcohols, Ethers, Phenols (con't)		Carboxylic Acids	
R–C(R')(OH)–**H**	3.3 - 4	R–C(=O)–O–**H**	10.5 - 12
R–O–C(R')(R")–**H**	3.3 - 4	R–C(R')(**H**)–C(=O)–O–H	2 - 2.6
C_6H_5–O–**H**	4 - 12	Aldehydes	
Amines, Amides		R–C(=O)–**H**	9-10
R–NH_2	1 - 5	Ketones	
R–C(=O)–N(R')–**H**	5 - 8	R–C(=O)–C(R')(R")–**H**	2 - 2.7
Esters			
R–C(R')(**H**)–C(=O)–O–R"	2 - 2.6		
R–C(=O)–O–C(R')(**H**)–R"	3.7 - 4.1		

Table 1 (con't) Chemical Shifts of Hydrogen Atoms in Various Functional Groups

Integration: The integration (the area under a signal or group of peaks) is proportional to the number of protons responsible for that signal or group of peaks. In the ^{1}H NMR spectrum of isopropylbenzene (Figure 2), the line marked "integral" can be used to calculate the relative areas of each set of peaks in the spectrum. The relative areas are determined by measuring (using a ruler) the vertical displacement (difference in mm) of the integral tracing on the right side of a signal or group of peaks to that on the left side of the signal or group of peaks. These displacements are 31 mm, 26, mm, and 5.5 mm for the groups of peaks having chemical shifts near $\delta = 7.2$, $\delta = 2.7$, and $\delta = 1.6$ respectively. The number of protons per molecule giving rise to each group of peaks can be calculated from the normalized (relative) areas of each set of peaks as shown in Table 2.

Table 2 Integration of Isopropylbenzene

Chemical Shift (δ)	Measured Displacement (mm)	Relative Area	# Protons
7.2	17.2	$\frac{(17.2)}{(3.6)} = 4.8$	5
2.7	3.6	$\frac{(3.6)}{(3.6)} = 1.0$	1
1.6	20.6	$\frac{(20.6)}{(3.6)} = 5.7$	6

Spin-Spin Splitting: Frequently the signal from single or equivalent protons appears as a symmetric group of regularly spaced peaks, as seen in the ^{1}H NMR spectrum of isopropylbenzene (Figure 2). This phenomenon is called spin-spin splitting. It is caused by interaction or coupling with nearby nuclei (normally protons). The distance (measured in Hz, not ppm) between these regularly spaced peaks is called the coupling constant. Modern NMR spectrometers can print peak lists in units of Hz and ppm. Hz and ppm can be interconverted as shown in equation 2.

$$\text{(value in Hz)} = \text{(value in ppm)} \bullet \text{(operating frequency of spectrometer in MHz)} \qquad (2)$$

In most cases the spin-spin splitting gives information about the number of protons on adjacent carbon atoms. Spin-spin splitting is not observed if the protons are equivalent (as in ethane, CH_3-CH_3), and it is not usually observed if the different protons are more than three bonds apart. The protons of hydroxyl groups (ROH) frequently do not show spin-spin splitting because of their rapid intermolecular exchange.

In isopropylbenzene, the methyl signal at $\delta = 1.6$ is split into two peaks of nearly equal intensity (called a "doublet"), and the CH signal at $\delta = 2.7$ is split into seven peaks (called a "septet") having relative intensities 1:6:15:20:15:6:1. The group of peaks near $\delta = 7.2$ is referred to by the more generic term "multiplet". We will return to the NMR spectrum of isopropylbenzene after discussing some simpler examples.

Example 1: Chemical Shifts and Integration

The ^{1}H NMR spectrum of 3-methyl-2-hydroxy-1-butyne is shown in Figure 3. This compound contains three different kinds of protons. Three peaks having chemical

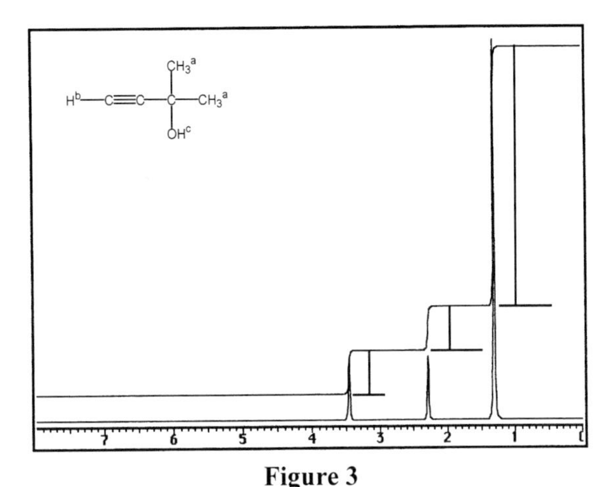

Figure 3

shifts of $\delta = 1.2$, $\delta = 2.3$, and $\delta = 3.4$ with relative areas of 6:1:1 respectively are observed. Based on their chemical shifts (see Table 1) and integration the peak at $\delta = 1.2$ must be due to the six methyl (H^a) protons, while the peak at $\delta = 2.3$ must be due to the acetylenic proton (H^b). By elimination the peak at $\delta = 3.4$ must be due to the OH proton (H^c). Each signal is a singlet; no spin-spin splitting is observed.

Example 2: Chemical Shifts, Integration, and Spin-Spin Splitting

The ^{1}H NMR spectrum of 1,1,2-tribromoethane is shown in Figure 4. Without considering spin-spin splitting, its NMR spectrum would be expected to have only two peaks of relative areas 1 and 2 corresponding to the one H^a and two H^b protons per molecule (see structure).

Br H^b

H^a—C—C—Br

Br H^b

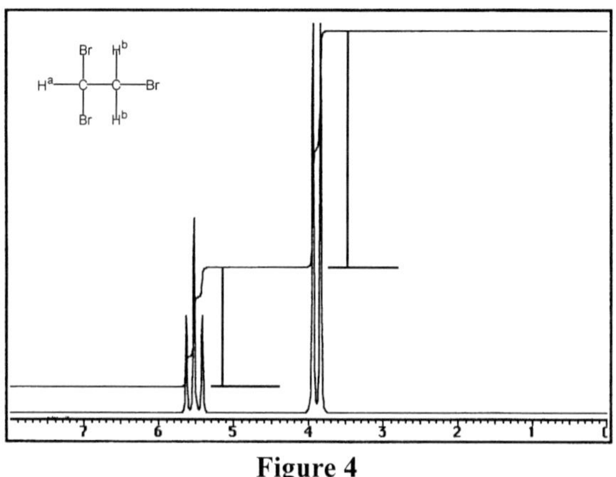

Figure 4

Instead of two peaks there are two groups of peaks, a "triplet" centered at δ = 5.5 (relative area 1) and a "doublet" centered at δ = 3.9 (relative area 2). The triplet and the doublet are examples of spin-spin splitting. The proton on the carbon with the two bromine atoms (H^a) would be expected to have a larger (more downfield) chemical shift than the protons on the carbon with one bromine atom (H^b). So the triplet centered at δ = 5.5 is assigned to the H^a proton, and the doublet centered at δ = 3.9 is assigned to the H^b protons. The integration also supports these assignments.

In discussing spin-spin splitting, the two energy states (spin states, defined earlier) of each proton are represented by two orientations of their magnetic moments, one aligned with the magnetic field (lower energy state, represented as ↑) and one aligned against the magnetic field (higher energy state, represented as ↓). The splitting occurs when the spin states from different kinds of neighboring protons influence one another (couple). This effect occurs via the electrons (a through-bond effect).

The energy difference between the two spin states is small enough that they are populated in nearly equal amounts at room temperature. Actually there is a very small excess in the lower energy state at equilibrium. Thus for any given proton in a compound, roughly half of the molecules will have that proton in the ↑ state and half of the molecules will have that proton in the ↓ state. For two equivalent protons in a molecule (e.g. H^b in 1,1,2-tribromoethane, Figure 4), the possible arrangements of their spin states are ↑↑, ↑↓, ↓↑, and ↓↓. The ↑↓ and ↓↑ states have the same energy. Statistically in a large ensemble of molecules the ↑↑ state, the ↑↓ and ↓↑ states, and the ↓↓ state will be present in a ratio of 1:2:1. So the NMR signal of a proton or protons (e.g. H^a) which couple with the H^b protons will be a triplet in a 1:2:1 ratio (as seen in the spectrum of 1,1,2-tribromoethane in Figure 4).

The coupling of protons is reciprocal. Since the signal of H^a is split because of the H^b protons, the signal of the H^b protons will be split because of the H^a proton. In a large ensemble of molecules roughly half of the molecules will have their H^a proton in the ↑

state and roughly half of the molecules will have their H^a proton in the ↓ state (a 1:1 ratio). So the NMR signal of the H^b protons will be a doublet with a ratio of 1:1 (as seen in the spectrum of 1,1,2-tribromoethane in Figure 4). Both groups of coupled protons will have the same coupling constant (the distance in Hz between the peaks of these regularly spaced multiplets). This observation can be useful in peak assignments.

As the number of identical protons increases, the number of possible arrangements of their spin states increases as shown in Table 3. The relative intensities follow Pascal's triangle. In general, a group of identical protons containing n protons will have n + 1 different spin states. So protons adjacent to carbon atoms having n equivalent protons will have their NMR signal split into n + 1 peaks (the n + 1 rule).

Table 3 Spin-Spin Splitting Patterns

# Equivalent Protons	Possible Arrangements of Spins Note: Arrangements within ( ) have the same energy	# Different Total Spin States	Relative Intensities
1	(↑) (↓)	2 (doublet)	1:1
2	(↑↑) (↑↓ ↓↑) (↓↓)	3 (triplet)	1:2:1
3	(↑↑↑) (↑↑↓ ↑↓↑ ↓↑↑) (↑↓↓ ↓↑↓ ↓↓↑) (↓↓↓)	4 (quartet)	1:3:3:1
4	(↑↑↑↑) (↑↑↑↓ ↑↑↓↑ ↑↓↑↑ ↓↑↑↑) (↑↑↓↓ ↑↓↑↓ ↑↓↓↑ ↓↑↓↑ ↓↓↑↑ ↓↑↑↓) (↓↓↓↑ ↓↓↑↑ ↓↑↓↓ ↑↓↓↓) (↓↓↓↓)	5 (quintet)	1:4:6:4:1
5	(↑↑↑↑↑) (↑↑↑↑↓ ↑↑↑↓↑ ↑↑↓↑↑ ↑↓↑↑↑ ↓↑↑↑↑) (↑↑↑↓↓ ↑↑↓↑↓ ↑↓↑↑↓ ↓↑↑↑↓ ↑↑↓↓↑ ↑↓↑↓↑ ↓↑↑↓↑ ↑↓↓↑↑ ↓↑↓↑↑ ↓↓↑↑↑) (↓↓↓↑↑ ↓↓↑↓↑ ↓↑↓↓↑ ↑↓↓↓↑ ↓↓↑↑↓ ↓↑↓↑↓ ↑↓↓↑↓ ↓↑↑↓↓ ↑↓↑↓↓ ↑↑↓↓↓) (↓↓↓↓↑ ↓↓↓↑↓ ↓↓↑↓↓ ↓↑↓↓↓ ↑↓↓↓↓) (↓↓↓↓↓)	6 (sextet)	1:5:10:10:5:1

Example 3: Chemical Shifts, Integration, and More Complicated Spin-Spin Splitting

The ^1H NMR spectrum of 1-butanal is shown in Figure 5. Specific regions of the spectrum have been expanded so the spin-spin splitting patterns can be seen more clearly.

1-Butanal contains four different types of protons. The chemical shifts, relative areas, and spin-spin splitting patterns of each of the four groups of signals are summarized in Table 4.

Table 4 ^{1}H NMR Spectrum of 1-Butanal

Chemical Shift (δ ppm)	Relative Area	Observed Spin-Spin Splitting Pattern
9.7	1	triplet
2.4	2	triplet of doublets
1.4	2	sextet
1.2	3	triplet

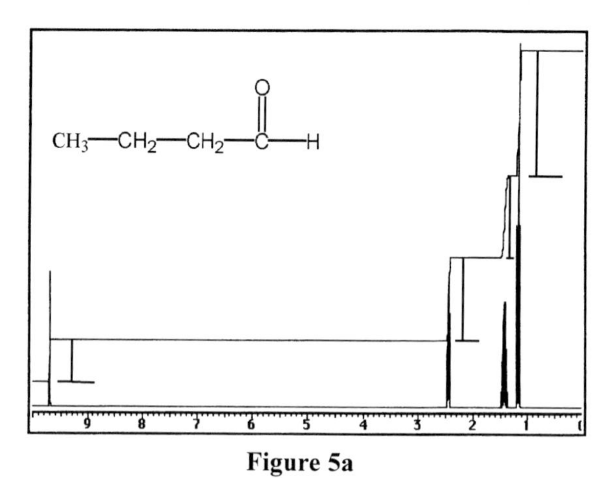

Figure 5a

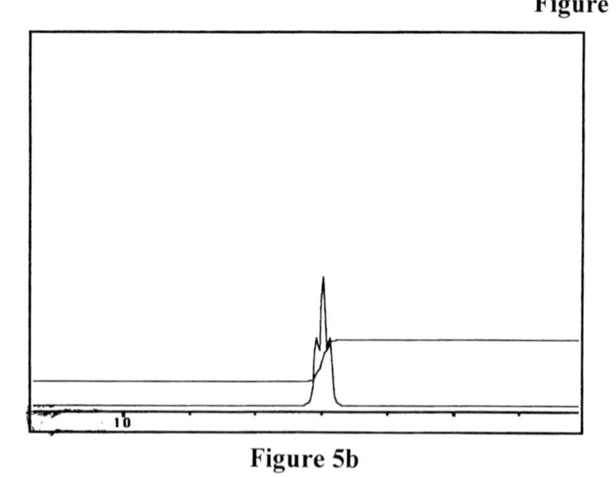

Figure 5b

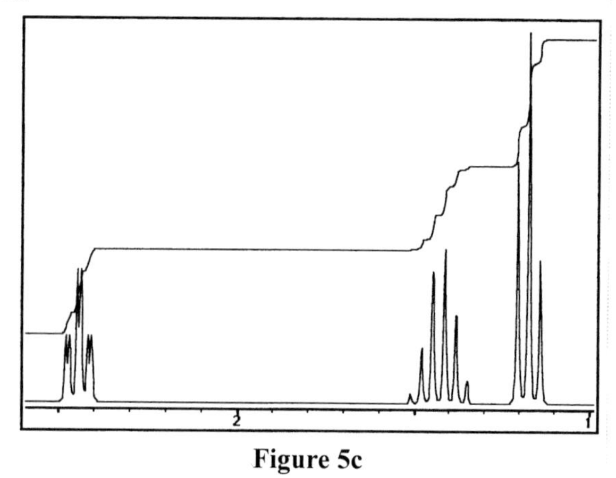

Figure 5c

Based on its relative area of 1 and its chemical shift, the signal at δ = 9.7 must be from the aldehyde proton (C-1 proton). Its signal is split into a triplet by the two equivalent protons attached to C-2.

Based on its relative area of 3 and its chemical shift, the signal at δ = 1.2 must be from the three equivalent methyl protons attached to C-4. The signal of these methyl protons is split into a triplet by the two equivalent protons attached to C-3.

The remaining two groups of signals in the spectrum both have relative areas of 2 and so must be from the two equivalent protons attached to C-2 and to C-3. The protons attached to C-2 are much closer to the electron withdrawing aldehyde carbonyl (C=O) group so their signal would be expected to shifted downfield relative to the signal of the two equivalent protons attached to C-3. So the sextet at δ = 1.4 can be assigned to the two equivalent protons attached to C-3. The protons attached to C-3 are adjacent to both the three equivalent protons attached to C-4 and to the two equivalent protons attached to C-2, so they are split by 5 (3 + 2) protons and thus appear as a sextet.

By elimination the resonance at δ = 2.4 must be due to the two equivalent protons attached to C-2. The splitting of this signal is more complicated. The signal appears to be a triplet in which each peak of the triplet is split into a doublet. The coupling constants (separations in Hz between these evenly spaced peaks) can be used to determine which protons are coupled to which, as mentioned above. The coupling constants associated with the sextet (C-3 protons, δ = 1.4), the three peaks of the triplet at δ = 1.2 (C-4 protons), and the three peaks of the triplet part of this triplet of doublets (C-2 protons, δ = 2.4) are all the about same (7 Hz), while those associated with the two peaks of each doublet of this triplet of doublets (C-2 protons) as well as the three peaks of the triplet at δ = 9.6 (C-1 proton) are much smaller (about 2 Hz). Thus the signal of the C-2 protons is split into a triplet by the two equivalent C-3 protons and each component of this triplet is split into a doublet by the single aldehyde (C-1) proton.

Based on the above discussion the ^{1}H-NMR spectrum of isopropylbenzene can now be completely assigned. The doublet (relative area 6) centered at δ = 1.6 is assigned to the six equivalent methyl protons split by the single benzylic proton. The septet (relative area 1) at δ = 1.6 is assigned to the single benzylic proton that is split by the six equivalent methyl protons. The group of peaks near δ = 7.2 must be due to the five aromatic protons in the molecule. Although these protons are not absolutely equivalent, their chemical shifts are very similar since the electronic effect of the substituent (alkyl group) is very small. For protons which have very similar chemical shifts, the n + 1 rule breaks down, and coupling constants may not be able to be simply measured by the spacings between the peaks.

Spin-Spin Splitting in Aromatic Rings Aromatic ring protons usually occur about δ = 6.0-8.5 in NMR spectra. In some cases, for example in isopropylbenzene discussed above, the chemical shifts of the aromatic protons are very similar, and the typical spin-spin splitting patterns are not observed. However substituents which are strongly electron donating or electron withdrawing can cause the aromatic protons to be in sufficiently different environments that their signals are more separated, and simple splitting patterns for some or all of the protons can be observed. Long-range couplings through the aromatic rind can further split these multiplets by a small amount. These patterns can aid in structural assignments.

Monosubstituted benzenes. Considering only ortho coupling (coupling between protons on adjacent carbons on the aromatic ring), the n+1 rule predicts that the aromatic protons of a monosubstituted benzene appear as a doublet (ortho protons), triplet (para protons) and a triplet(or doublet of doublet) (meta protons). With nitrobenzene (Figure 6) it is possible to see the doublet (H^a), the triplet (H^c), and the triplet (or doublet of doublets) (H^b). However, in many monosubstituted benzenes, such as toluene (Figure 7), the chemical shifts of the aromatic protons are very similar. In these cases, sometimes not all of the splittings are visible, and sometimes the splitting patterns can become rather complicated.

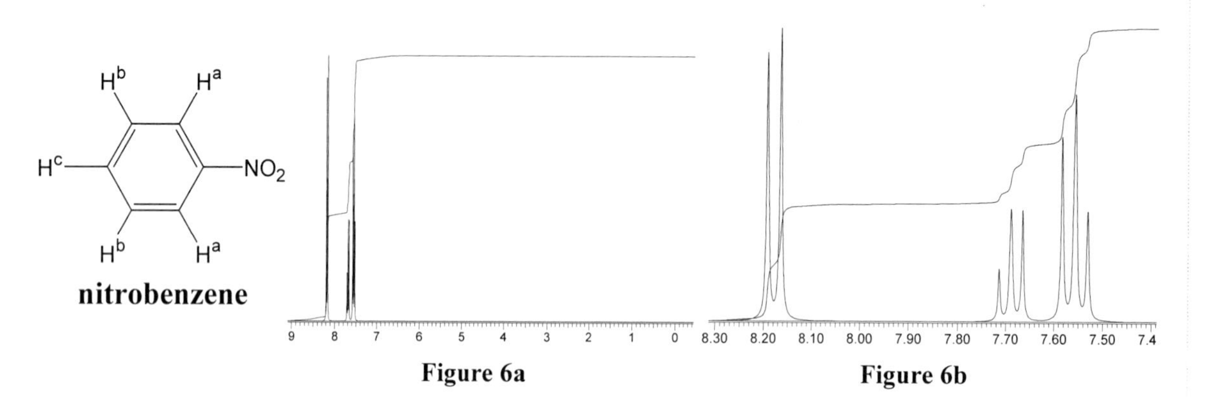

Figure 6a

Figure 6b

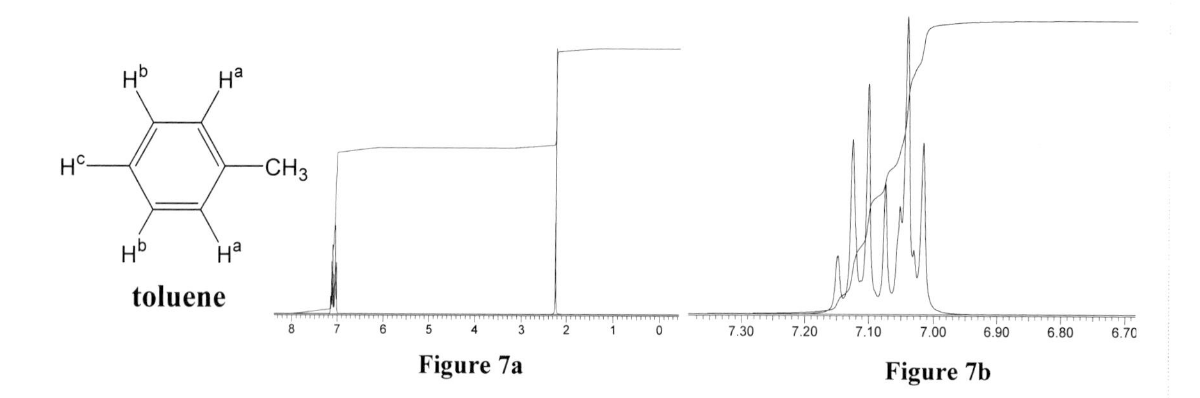

Figure 7a

Figure 7b

Disubstituted benzenes. The aromatic protons in para-disubstituted benzenes frequently appear as two doublets as expected. For example in p-bromotoluene (Figure 8), H^a and H^b each appear as a doublet. The aromatic protons in many ortho-disubstituted benzenes appear as two doublets and two triplets. For example in acetylsalicylic acid (aspirin, Figure 9), H^a and H^d each appear as a doublet and H^b and H^c each appear as a triplet. The aromatic protons in meta-disubstituted benzenes can appear as a singlet, two doublets, and a triplet as predicted from the n+1 rule. For example, in 3-nitrobenzyl bromide (Figure 10), H^a appears as a singlet, H^b and H^c each appear as doublets, and H^d appears as a triplet. As with the monosubstituted benzenes, if the chemical shifts of the aromatic protons in a disubstituted benzene are too similar (as in m-xylene, Figure 11), sometimes not all of the splittings are visible, and sometimes the splitting patterns can become rather complicated.

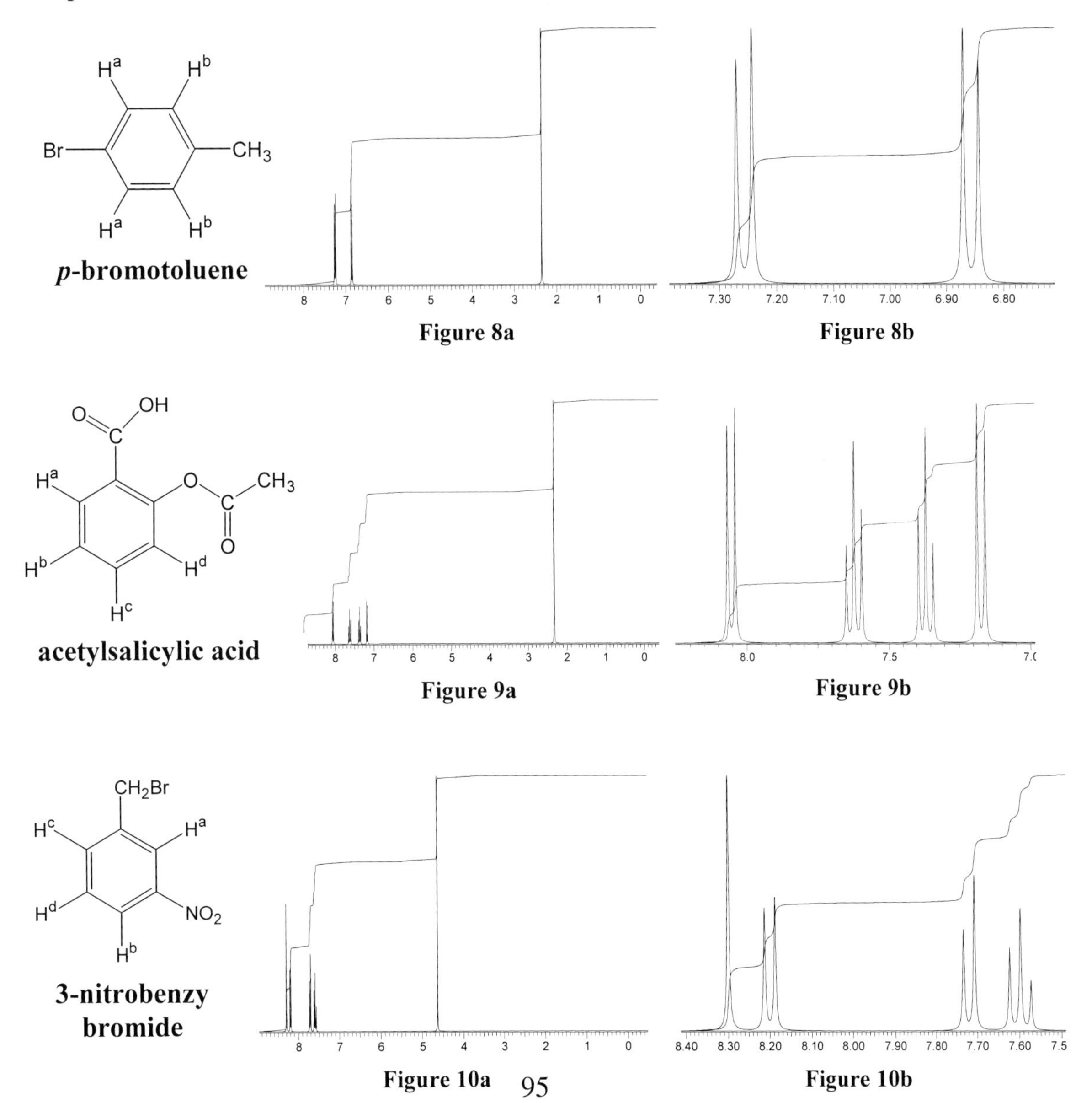

Figure 8a

Figure 8b

Figure 9a

Figure 9b

Figure 10a

Figure 10b

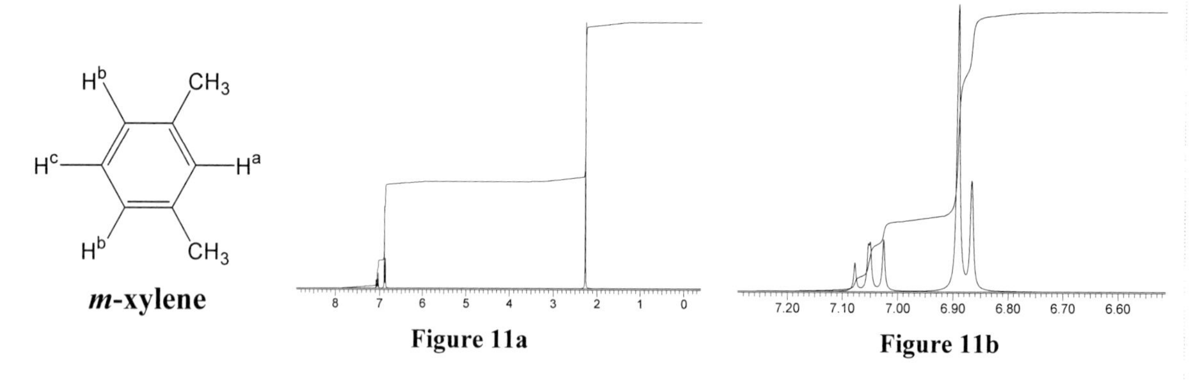

***m*-xylene**

Figure 11a

Figure 11b

Problems

1. How many different kinds of protons (H atoms) are there in each of the following molecules?

   A. CH_3CH_3
   B. $CH_3CH_2CH_3$
   C. $(CH_3)_2CHCH_2CH_3$
   D. $H_2C{=}CH_2$
   E. $CH_3CH{=}CH_2$
   F. $C_6H_5NO_2$
   G. $C_6H_5CH_3$

   Answers: A (1), B (2), C (4), D(1), E(4), F(3), G(4)

2. What is the expected splitting pattern and relative area for each of the two signals expected from the following compound?

$$\mathrm{Cl{-}CH^{a}(Cl){-}CH_3{}^{b}}$$

Answer: Using the n+1 splitting rule of spin-spin coupling the one H^a proton splits the signal of the H^b protons into a doublet of relative area 3 and the three H^b protons split the signal of the H^a proton into a quartet of relative area 1.

3. What are the expected splitting patterns and relative areas for the signals of the H^a and H^b protons in the following molecule?

$$
\begin{array}{c}
CH_3{}^a \\
| \\
H_3{}^aC - C - H^b \\
| \\
CH_3{}^a
\end{array}
$$

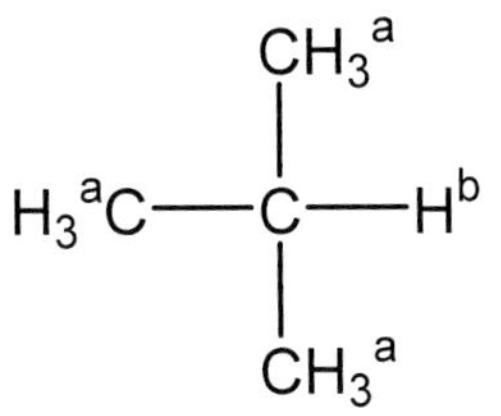

Answer: The signal of the H^a protons is split into a doublet and has relative area of 9. The signal of the H^b proton is split into a multiplet and has relative area 1.

4. What are the expected spin-spin splitting patterns (multiplicities) and relative areas for the signals of the H^a, H^b, and H^c protons in the following molecule?

$$CH_3{}^a - CH_2{}^b - CH_2{}^c - NO_2$$

Answer: The signals of the H^a and H^c protons are each split into a triplet by the two H^b protons. These signals will have relative areas of 3 and 2, respectively. The signal for the H^b protons will be split into a sextet by the five ($3H^a + 2H^c$) protons and will have a relative area of 2.

5. Match each of the following structures with its ^{1}H-NMR spectrum. The structures are numbered a-f and the spectra are numbered 5-1 to 5-6.

Problem 5					
	Compound Structure	Spectrum Number		Compound Structure	Spectrum Number
a	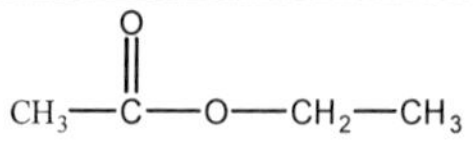 $CH_3-C(=O)-O-CH_2-CH_3$	________	d	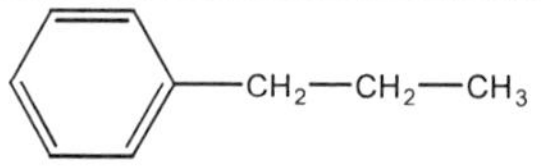 $C_6H_5-CH_2-CH_2-CH_3$	________
b	$CH_3-CH_2-C_6H_4-I$	________	e	CH_3-CH_2-I	________
c	$CH_3-C(=O)-O-CH(CH_3)-CH_3$	________	f	$C_6H_5-C(=O)-O-CH_3$	________

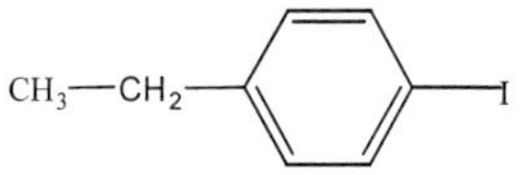

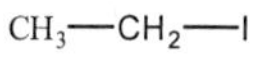

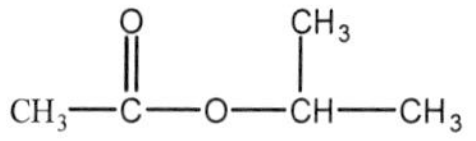

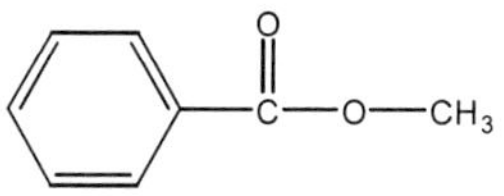

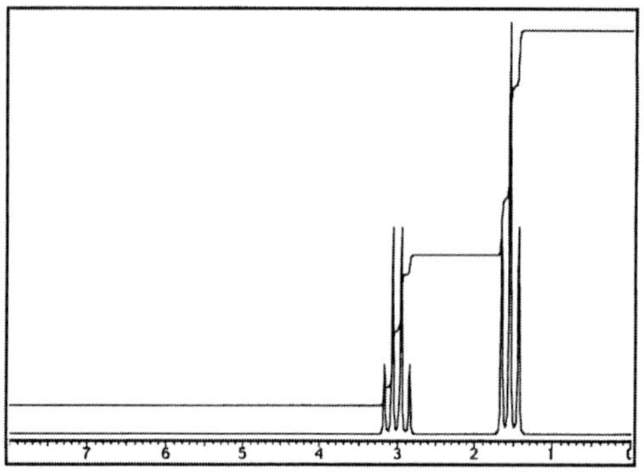

NMR Spectrum 5-1

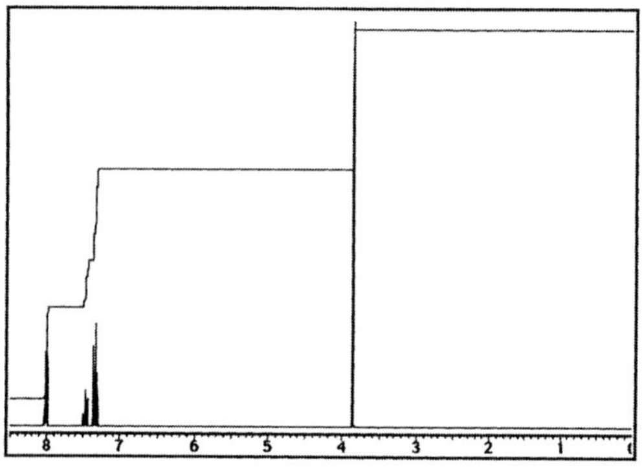

NMR Spectrum 5-2

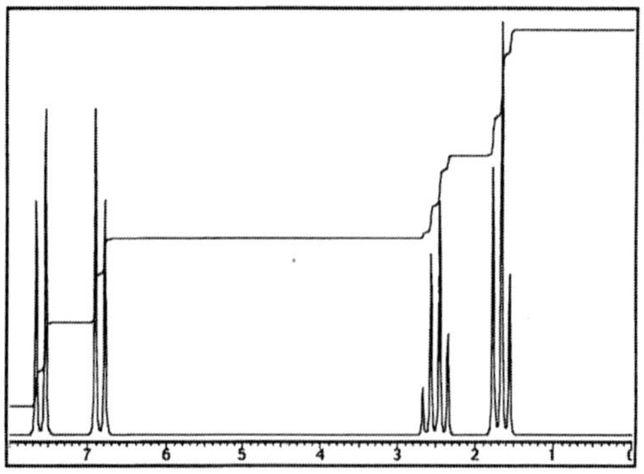

NMR Spectrum 5-3

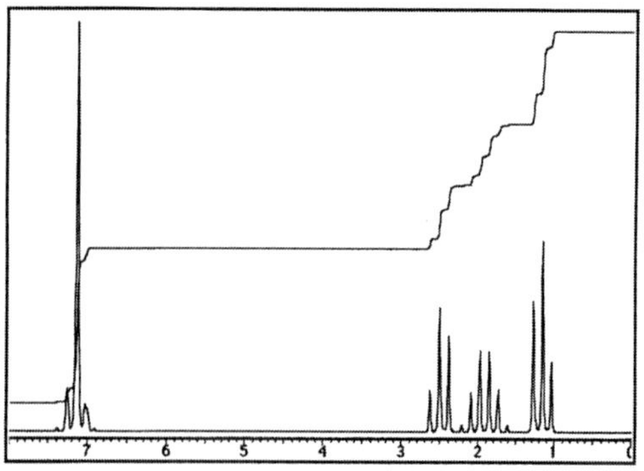

NMR Spectrum 5-4

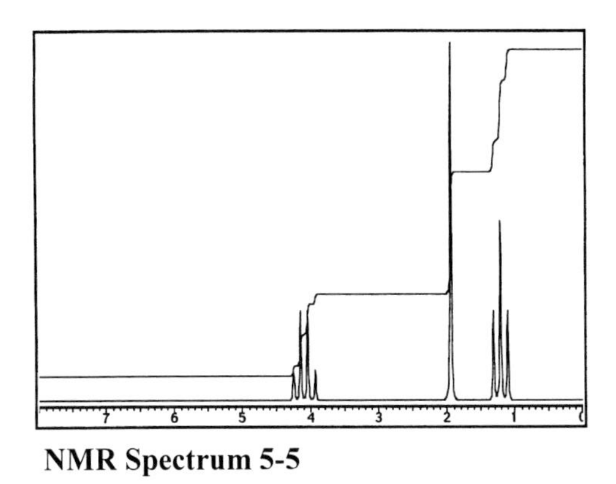

NMR Spectrum 5-5

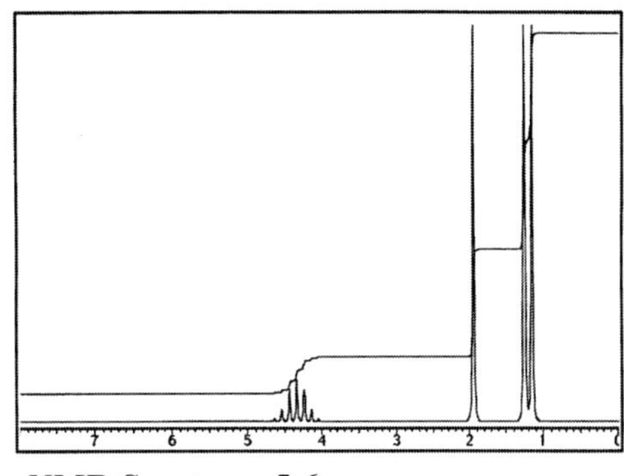

NMR Spectrum 5-6

6. A compound has molecular formula $C_4H_{10}O$. Its ^{1}H-NMR spectrum is shown below. What is the structure of this compound? Assign all the signals and explain the observed areas and splitting patterns in its ^{1}H-NMR spectrum.

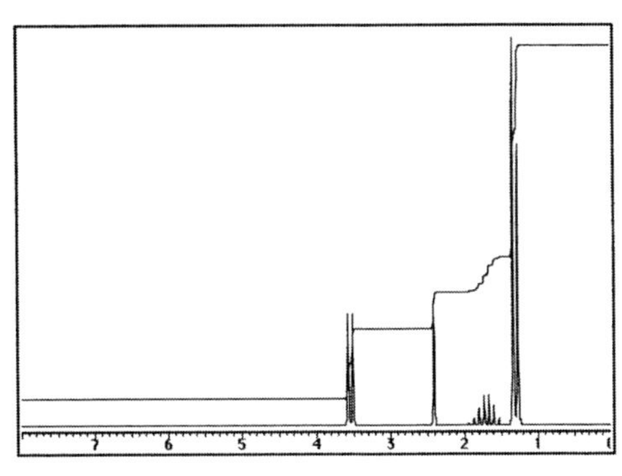

Problem 6

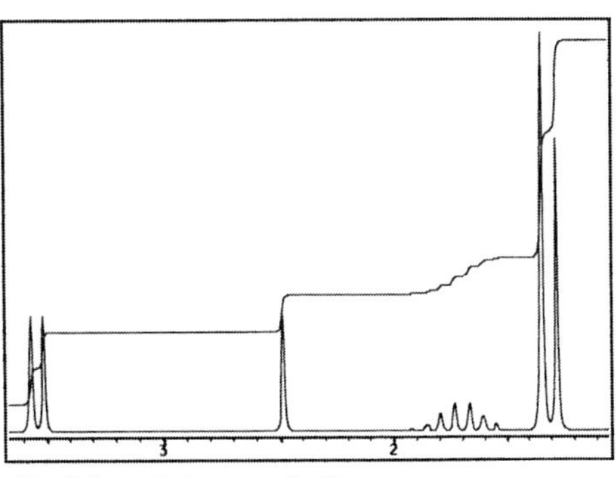

Problem 6 (expanded)

MASS SPECTROMETRY

Mass spectrometry provides information on the mass (MW) of a molecule and its fragments. From this information we can determine the molecular weight of the molecule and considerable information about its structure.

The most common type of mass spectrometry is the detection (and measurement of abundance) of positive ions. These ions are formed by the attack of high energy electrons (usually 70 eV) upon a sample. The **molecular ion** (M^+) is formed by loss of an electron, and various fragment (or rearrangement) ions are also formed, depending on the structure.

$$M \quad + \quad e^- \,(70 \text{ eV}) \quad \rightarrow \quad M^+ \quad + \quad 2\, e^-$$

By piecing together these fragment (and/or rearrangement) ions, a mass spectroscopist can figure out (sometimes) the structure of the starting compound.

Mass spectrometry assumes that the sample can be made volatile (usually by some ancillary technique such as gas or liquid chromatography) and that the fragment (or rearrangement) ions can be separated from each other. Separation of the ions is usually accomplished by using a magnetic sector mass spectrometer, a quadrapole mass spectrometer, or a time-of-flight mass spectrometer.

The resulting mass spectrum is usually presented as a graph of relative intensity (abundance of an ion) versus its mass to charge (m/z) ratio. The largest peak is designated as the **base peak** (relative intensity 100%) and is scaled against every other peak. The molecular ion (M^+) may or may not be present depending on the type of compound being examined. For example, alcohols sometimes lose H_2O (mass 18) and then do not show a molecular ion. With experience, however, the loss of water can be correlated with the presence of a small C_nH_{2n} ion in the mass spectrum to suggest the presence of an aliphatic alcohol.

For the most part, mass spectrometry does not provide information about stereoisomers (and certainly not about enantiomers) including Z and E isomers, or even positional isomers. For instance, the differences in the mass spectra of the following positional isomers $CH_3(CH_2)_3CO_2(CH_2)_4CH_3$, $CH_3CH_2CH(CH_3)CO_2(CH_2)_4CH_3$, and $CH_3CH(CH_3)CH_2CO_2(CH_2)_4CH_3$ are minute, and authentic samples must be used along with gas chromatographic retention times for positive identification. However the position of a ketone group along an alkyl chain can frequently be determined by mass spectrometry. Methyl ketones exhibit an abundant ion (at times it is the base peak) at m/z = 43 (CH_3CO^+). A methyl ketone with the structure shown below will also have a second peak at m/z = 58 (due to a McLafferty rearrangement) as long as there is at least one H on the γ carbon (see scheme below).

$$CH_3-\underset{}{C}(=O)-\underset{\alpha}{CH_2}-\underset{\beta}{CH_2}-\underset{\gamma}{CH(H)}-R \longrightarrow CH_3-C(OH)=CH_2^{+} + CH_2=CH-R$$

m/z = 58

For example the mass spectrum of 2-dodecanone (see Figure) has a large peak at m/z = 43 (CH_3CO^+) , and a McLafferty rearrangement ion at m/z = 58 is the base peak. A

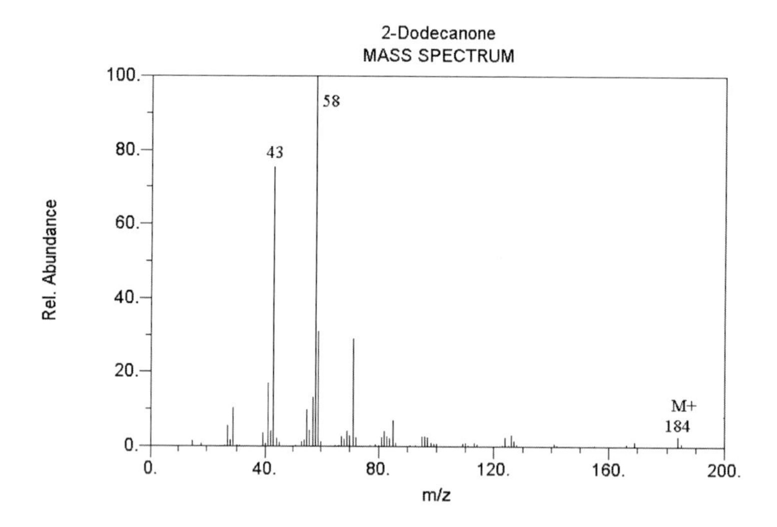

molecular ion is observed at m/z = 184. Peaks due to loss of methyl (M - 15 = 169) and loss of water (M - 18 = 166) are also observed. In an ethyl ketone the original fragment ion occurs at m/z = 57 ($CH_3CH_2CO^+$) and the McLafferty rearrangement ions occur at m/z = 72 or higher. This rearrangement peak moves in increments of 14 AMU (atomic mass units, CH_2) towards higher mass. If the structure of the molecule allows, a McLafferty rearrangement can occur in both directions (see scheme below). Only a double (or triple) bond is necessary along with an available γ hydrogen. Substitution at the α position moves this rearrangement peak to higher mass.

$CH_3-CH(CH_2H)-CH_2-C(=O)-O-CH_2-CH(H)-CH(CH_3)-CH_3$

m/z = 172

$[CH_2=C(OH)-O-CH_2-CH(H)-CH(CH_3)-CH_3]^+$ + $[CH_3-CH(CH_2H)-CH_2-C(=O)-OH]^+$

m/z = 130 m/z = 102

+ $CH_3-CH=CH_2$ + $CH_2=CH-CH(CH_3)-CH_3$

Thus the mass spectrum of 6-dodecanone exhibits a molecular ion at m/z = 184 and much less intense McLafferty rearrangement ions at m/z = 114 and m/z = 128 than its isomer 2-dodecanone. A fragment ion is observed at m/z = 99 ($CH_3(CH_2)_4CO^+$) but no corresponding fragment ion ($CH_3(CH_2)_5CO^+$) is observed at m/z = 113. The base peak at m/z = 43 is $C_3H_7^+$.

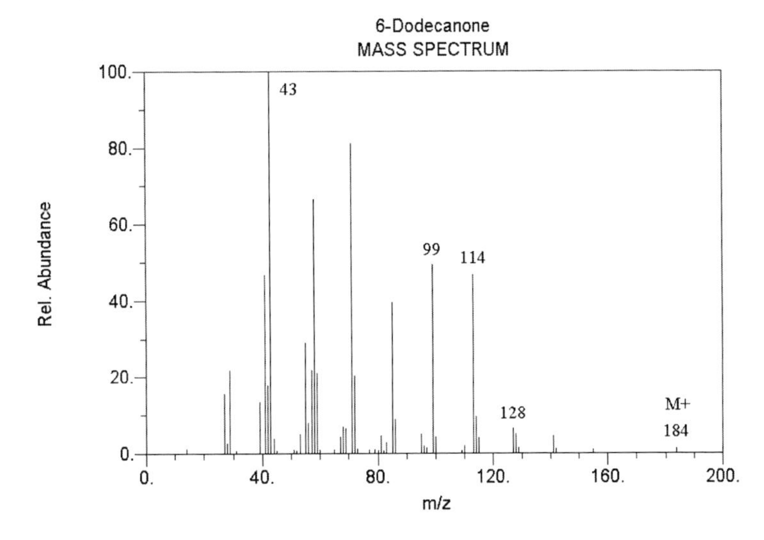

Hexylbenzene (see Figure) exhibits a molecular ion (M^+) at m/z = 162, a benzyl ion at m/z = 91 (base peak) and a McLafferty rearrangement peak at m/z = 92. The other fragment ions are very weak.

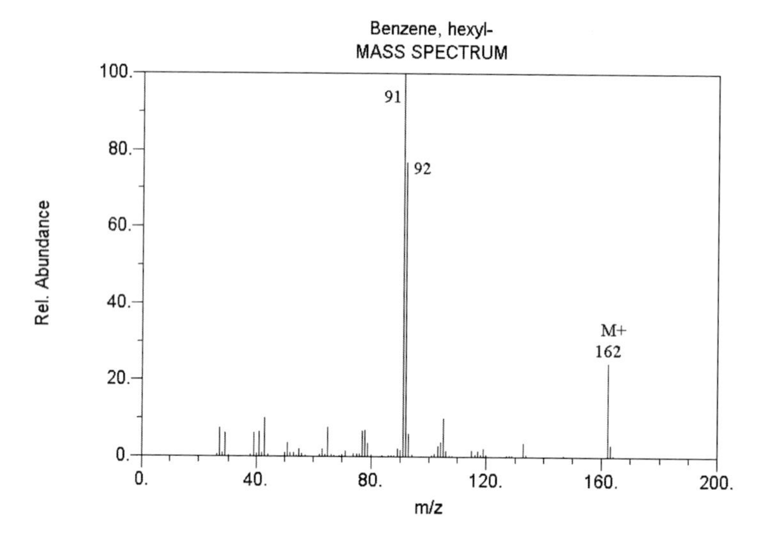

The mass spectrum of dodecane (see Figure) is typical of a long chain alkane exhibiting a base peak at m/z = 57 ($C_4H_7^+$) with higher mass ions of much lower intensity. The peaks at m/z = 43, 71, and 85 are $C_nH_{2n+1}^+$ ions with n = 3, 5, 6. A molecular ion is seen at m/z = 170.

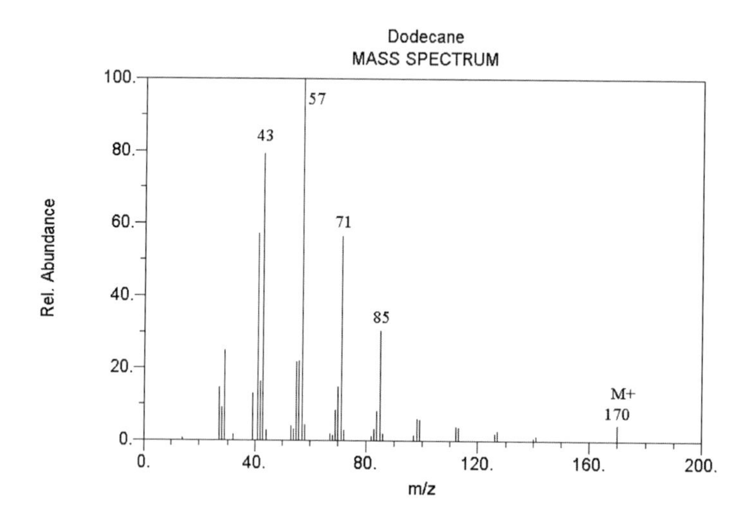

Branching of the alkyl chain, as in 2,2,4,6,6-pentamethylheptane (see Figure) gives a base peak at m/z = 57 ($(CH_3)_3C^+$), but the other fragment ions are much weaker than those of dodecane. With this compound a molecular ion at m/z = 170 is not observed. The highest mass peak is at m/z = 155 (loss of 15, CH_3).

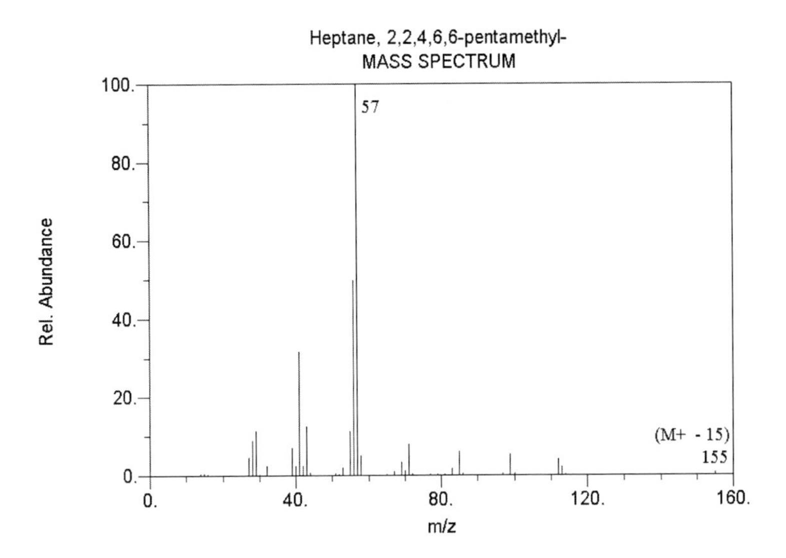

1-Dodecene (see Figure) has many of the same ions as dodecane. However the corresponding ions with two less hydrogens ($C_nH_{2n-1}^+$, m/z = 41, 55, 69, 83, 97) are the most intense. A molecular ion is observed at m/z = 168 and the ion at m/z = 140 is due to mass loss of 28 (C_2H_4).

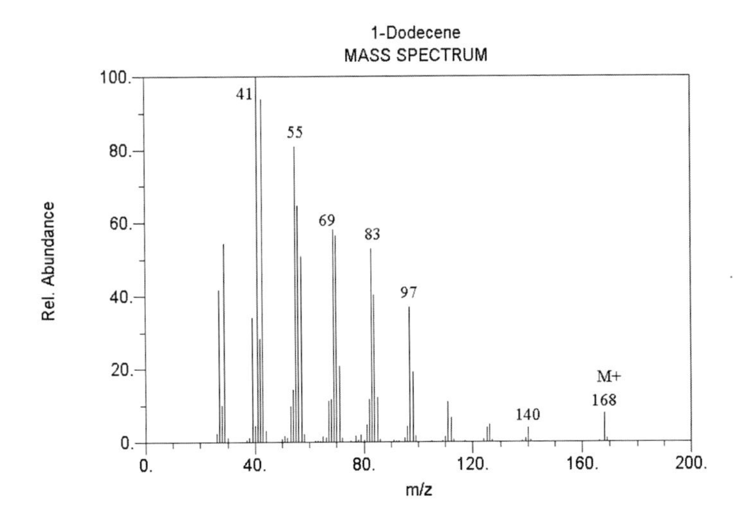

1-Dodecanol (see Figure) has essentially the same spectrum as 1-dodecene with the exception of weak ions at m/z = 45, 59, 73, and 87 corresponding to $C_nH_{2n+1}O^+$. With 1-dodecanol no molecular ion at m/z = 186 is observed, and the highest fragment ion at m/z = 168 is due to loss of water (mass 18) to give an olefinic fragment. The same series of peaks at m/z = 41, 55, 69, 83, 97 observed with 1-dodecene are present in the mass spectrum of 1-dodecanol.

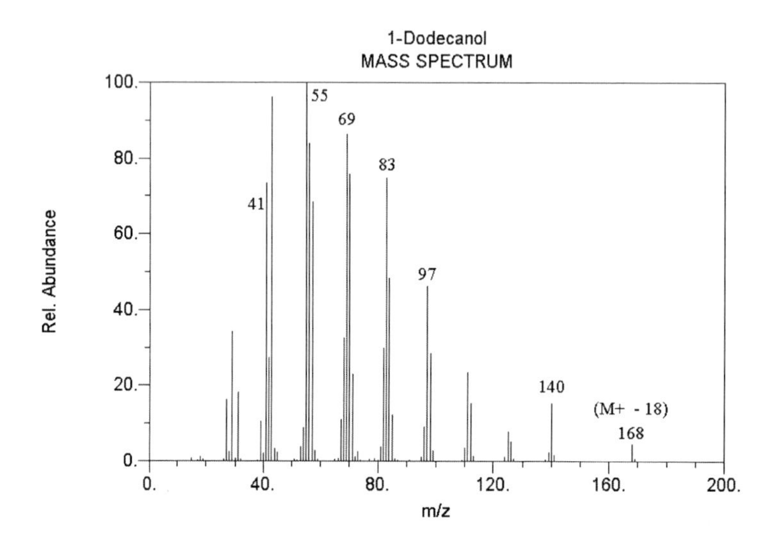

The spectrum of dodecanoic acid (see Figure) is dominated by a fragment ion at m/z = 73 (base peak), corresponding to γ-cleavage ($CH_2CH_2CO_2H^+$). A McLafferty rearrangement ion at m/z = 60 is also very prominent. The molecular ion is observed at m/z = 200.

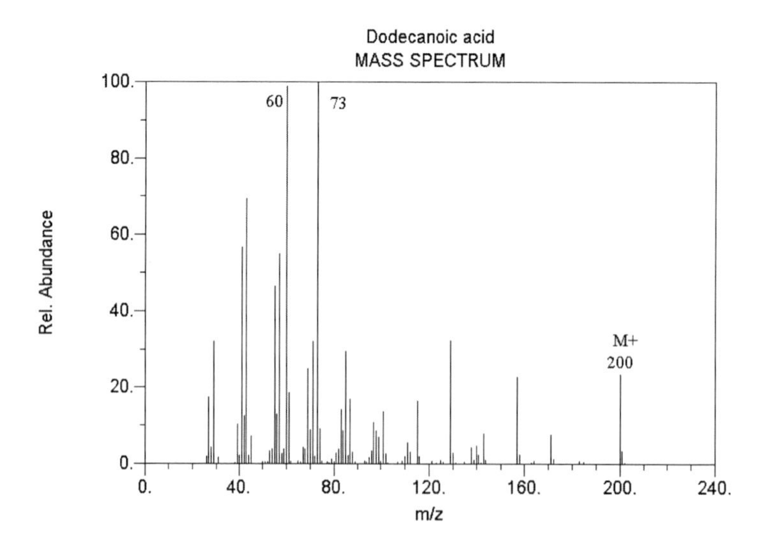

The spectrum of methyl undecanoate (see Figure) is also dominated by a McLafferty rearrangement ion (now at m/z = 74) and a γ-cleavage ion at m/z = 87 ($CH_2CH_2CO_2CH_3^+$). A molecular ion is again observed at m/z = 200.

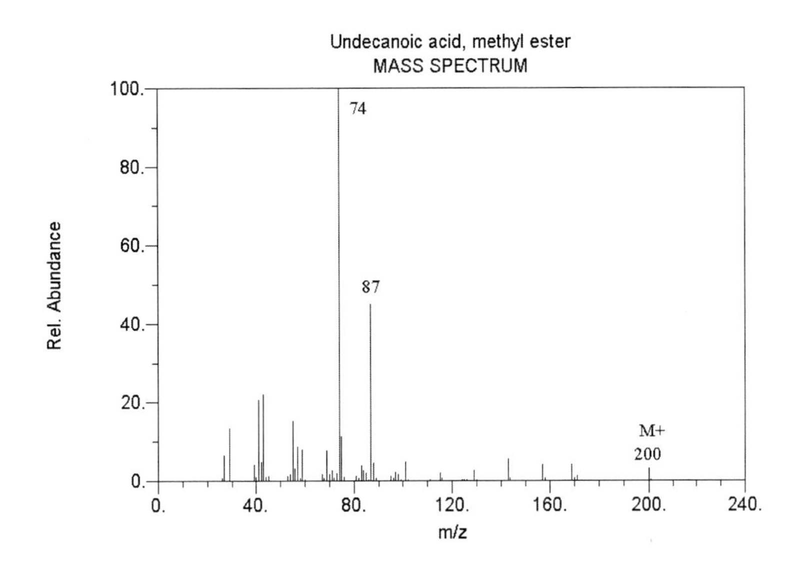

Other examples of structural information obtained from mass spectrometry include determining whether the compound contains Cl, Br, or N. Naturally occurring chlorine contains ^{35}Cl and ^{37}Cl in a ratio of 3:1 (see Table of Principal Stable Isotopes of Common Elements) so compounds containing one chlorine atom have a $(M+2)^+$ ion having 33% abundance compared to the M^+ molecular ion. Similarly bromine has two isotopes, ^{79}Br and ^{81}Br, of almost equal abundance, so compounds containing one bromine have a $(M+2)^+$ ion having 98% abundance compared to the M^+ ion. Compounds that contain an odd number of nitrogen atoms give a molecular ion (M^+) that has an odd m/z (the **nitrogen rule**), and ions that result from single cleavages have an even m/z.

The spectrum of 1-dodecanamine (see Figure) is dominated by a peak at m/z = 30 ($CH_2NH_2^+$) and exhibits a small molecular ion at m/z = 185. The other fragment ions are all weak. The spectrum of dodecanamide (see Figure) is again dominated by a McLafferty rearrangement ion at m/z = 59 and a γ-cleavage fragment at m/z = 72 ($CH_2CH_2CONH_2^+$). A small molecular ion is observed at m/z = 199. The other fragment ions are again quite small.

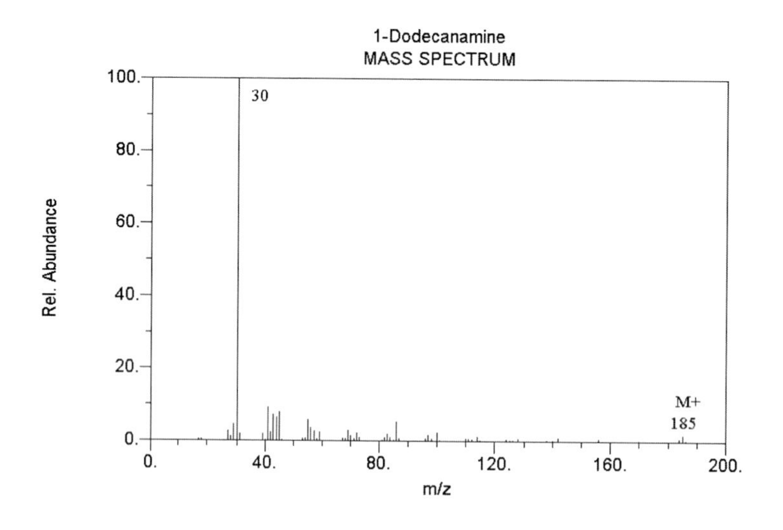

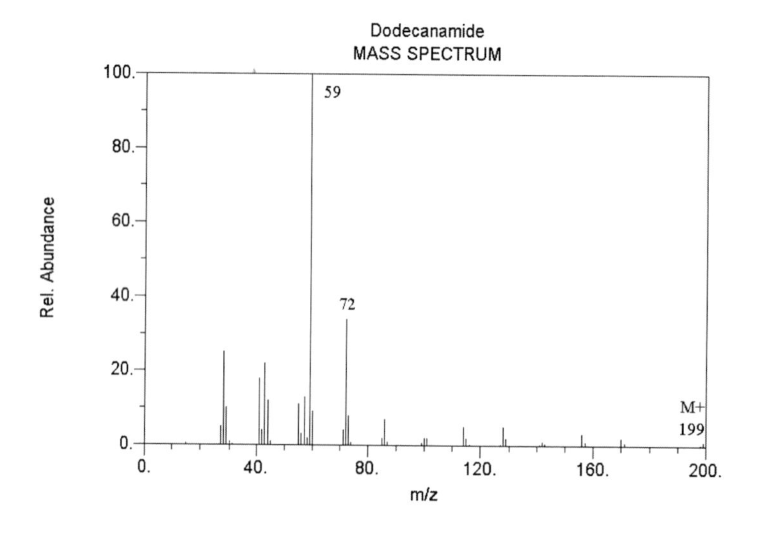

The spectrum of 1-bromododecane shows a molecular ion at m/z = 248 and a M+2 ion at 250 of almost equal intensity. Loss of C_7H_{15} (mass 99) also gives fragment ions at m/z = 149 and 151 of essentially equal intensity. Additional fragment ions containing bromine are observed at m/z = 135 and 137. These ions probably correspond to $(CH_2)_5Br^+$ and $(CH_2)_4Br^+$ respectively. These four ions also provide evidence that a bromine atom is in the molecule. The base peak at m/z = 57 and the intense peak at m/z = 43 correspond to $C_4H_9^+$ and $C_3H_7^+$ fragments respectively.

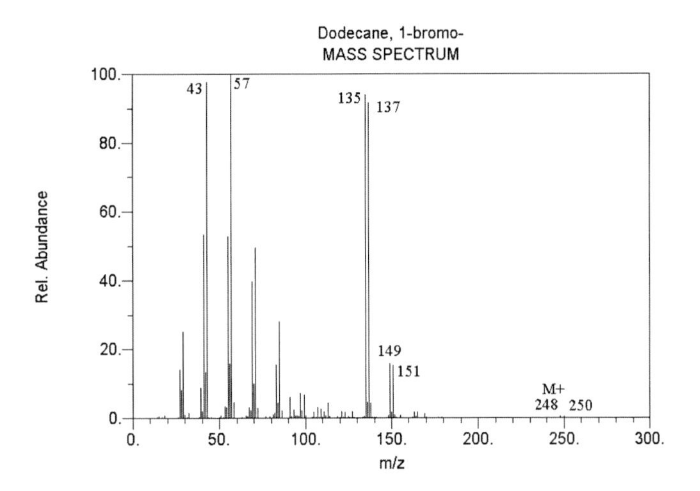

The spectrum of 1-chlorododecane (see Figure) does not exhibit a molecular ion at m/z = 204, and the highest fragment ion at m/z = 168 is due to loss of HCl (mass 36). Evidence of a chlorine atom being present in the molecule is the presence of fragment ions at m/z = 119 and 121, 105 and 107, and 91 and 93, with the relative intensities within each pair being around 3:1. These fragment ions correspond to $(CH_2)_6Cl^+$, $(CH_2)_5Cl^+$, and $(CH_2)_4Cl^+$ respectively.

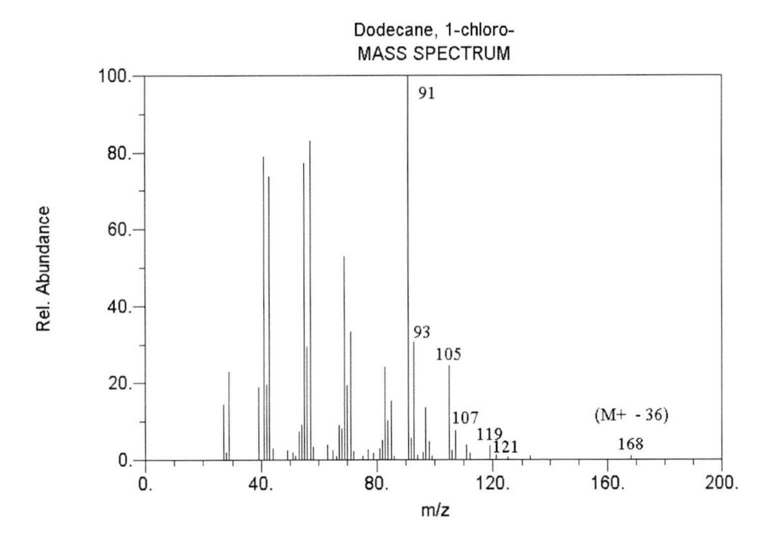

Lists of both common fragments and common losses (difference in mass between the molecular ion and the fragment) along with their respective mass or m/z's are given below. A table of the principal stable isotopes of the common elements found in organic compounds is also included below.

Common Losses Species	Mass	Common Fragments Ion	m/z
CH_3	15	$C_3H_5^+$	41
H_2O	18	$C_3H_7^+$	43
CH_2CH_2	28	CH_3CO^+	43
CO	28	CH_3CHOH^+	45
N_2	28	$CH_3CH_2CO^+$	57
CH_3CH_2	29	$C_4H_9^+$	57
$H_2O + CH_3$	33	$CH_3CO_2H_2^+$	61
^{35}Cl	35	$C_6H_5^+$	77
^{37}Cl	37	$^{79}Br^+$	79
$CH_3CH{=}CH_2$	42	$^{81}Br^+$	81
CH_3CHCH_3	43	$C_6H_5CH_2^+$	91
CH_3CO	43	$C_6H_5CH_2CH_2^+$	105
CH_3CO_2H	60	$C_6H_5CO^+$	105

Principal Stable Isotopes of Common Elements

Element	Most Common Isotope		Other Isotopes			
	M	% Abundance	M+1	% Abundance	M+2	% Abundance
C	^{12}C	98.890	^{13}C	1.108		
H	^{1}H	99.984	^{2}H	0.016		
O	^{16}O	99.76	^{17}O	0.04	^{18}O	0.20
N	^{14}N	99.64	^{15}N	0.36		
Cl	^{35}Cl	75.4			^{37}Cl	24.6
Br	^{79}Br	50.57			^{81}Br	49.43

Problems

1. Compounds A, B, and C are isomers ($C_{10}H_{12}O$). What are their structures?

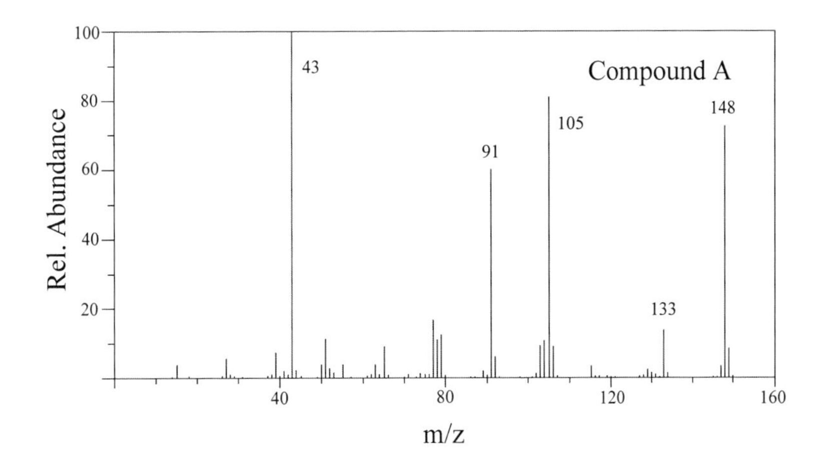

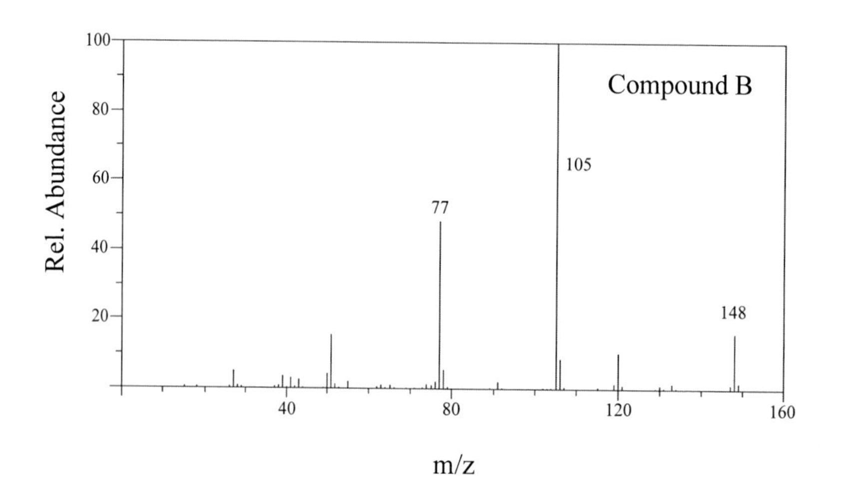
Compound B
Rel. Abundance
100
80
60
40
20
105
77
148
40
80
120
160
m/z

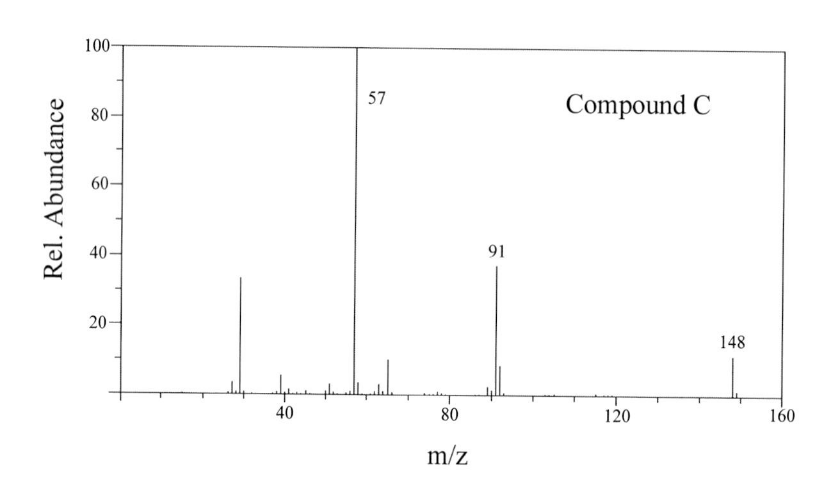
Compound C
Rel. Abundance
100
80
60
40
20
57
91
148
40
80
120
160
m/z

2. What is the structure of Compound D? Propose reasonable structures for the fragments having m/z of 43 and 57.

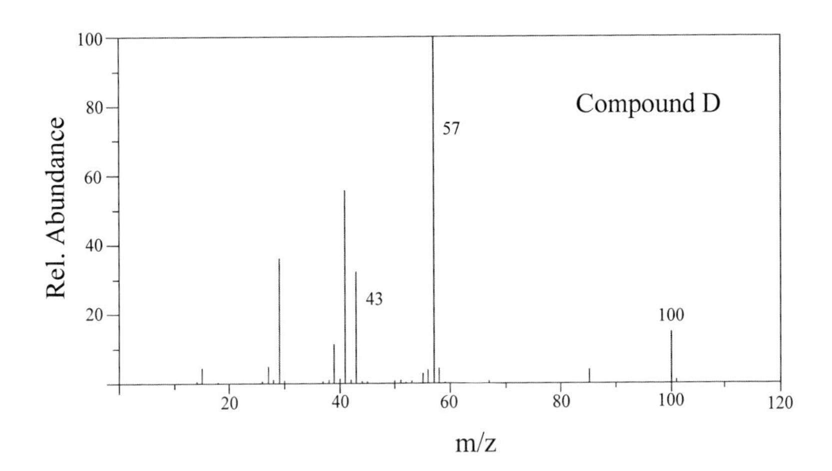

3. Compounds E and F are isomers. What are their structures? For compound E propose reasonable structures for the fragments having m/z of 57, 93, 95, 107, and 109. For compound F propose reasonable structures for the fragments having m/z of 57, 107, and 109.

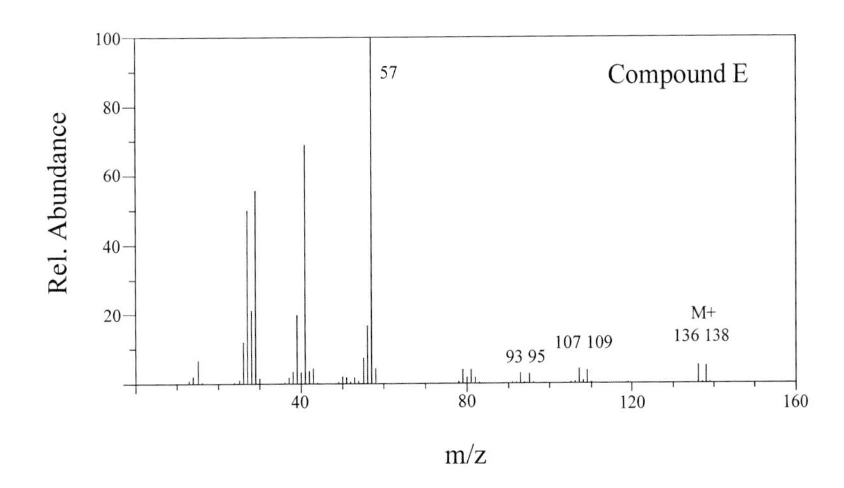

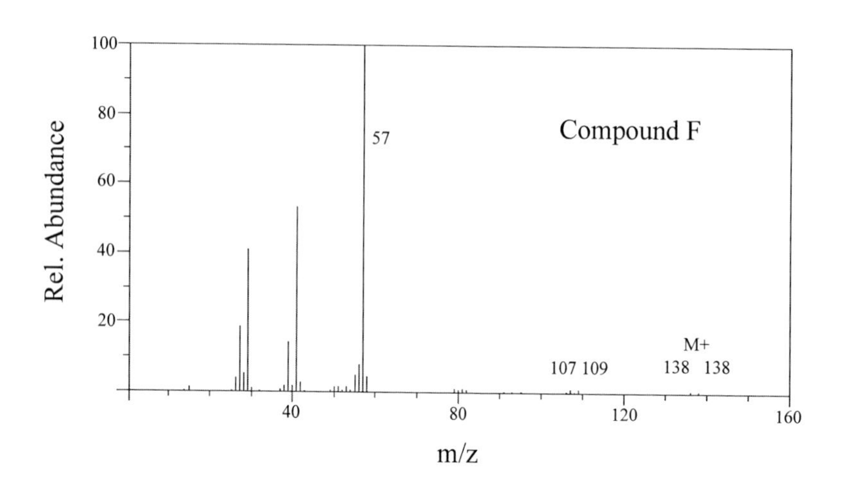

4. What is the structure of Compound G ($C_{19}H_{16}O$)? Propose reasonable structures for the fragments having m/z of 77, 106, 183, and 243?

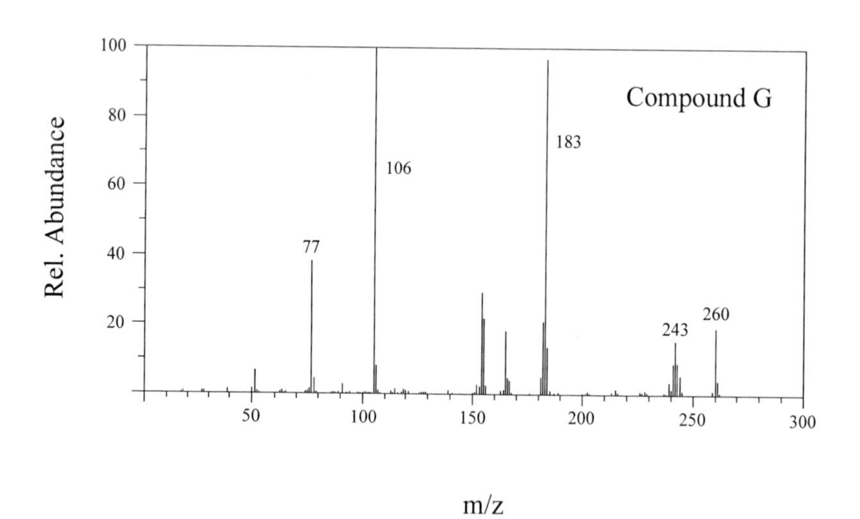

DISTILLATION

Distillation is used for ISOLATION and PURIFICATION of liquids.

heat   cool
liquid ----------> vapor ----------> liquid

Generally, the composition of the vapor is different from the liquid which is heated, and the composition of the cooled liquid (condensate) is the same as the vapor.

1. Pure liquids

A liquid in a distilling flask is heated with boiling chips (also called boiling stones or "Boileezers") or wooden sticks to prevent sudden eruptive boiling (bumping). If the cooled vapor is collected in the same flask, the process is called REFLUXING. Refluxing is useful for reactions in which there is prolonged heating of a low-boiling solvent, in order to prevent evaporation of the solvent. If the cooled vapor is collected in a different flask, the process is called DISTILLATION.

The BOILING POINT (bp) of a liquid is the temperature at which the vapor pressure of the liquid equals the external pressure.

vapor pressure = external pressure

At a given pressure, the bp of a pure compound is a characteristic PHYSICAL CONSTANT of the compound.

In general, for the same type of compounds, a compound having a low molecular weight will have a low boiling point and a high vapor pressure at room temperature, while a compound having a high molecular weight will have a high boiling point and a low vapor pressure at room temperature.

low molecular wt. = low bp = high vapor pressure at room temperature
high molecular wt. = high bp = low vapor pressure at room temp

A high temperature for distillation may be undesirable because the compound may decompose at the high temperature, or, in an industrial setting, the high cost of fuel could make the distillation cost ineffective.

Two alternatives to high temperature distillation are VACUUM DISTILLATION, where the external pressure is lowered by use of a vacuum pump, or STEAM

DISTILLATION, where the high-boiling liquid is co-distilled with an immiscible liquid, usually water. In a steam distillation, each of the IMMISCIBLE liquids exerts its own vapor pressure independently:

total vapor pressure	= vapor pressure organic	+ vapor pressure steam
760 torr near 100 °C	very low	very high

The DISTILLATE from a steam distillation will consist of a small amount of organic liquid and a large amount of water. The organic layer is easily separated from the water. A steam distillation will be used in the synthesis of cyclohexanone.

2. Solutions (of two or more compounds)

For "ideal solutions" the total vapor pressure depends on the vapor pressure (or bp) of the individual compounds and the amount of each compound present in the liquid mixture

Specifically, for a two-component system (containing liquids A and B)

$$P_{total} = P_A^o N_A + P_B^o N_B$$

where P_A^o is the vapor pressure of pure liquid A at a given temperature and N_A is the mole fraction of compound A in the solution where

$$N_A = \text{moles of A} / (\text{moles A} + \text{moles B})$$

N_A represents the amount of A in the LIQUID but $P_A^o N_A$ represents the amount of A in the VAPOR. P_A is the partial pressure of A in the vapor.

$$P_A = P_A^o N_A$$

From this equation, the compound with the HIGHER vapor pressure (the one with the LOWER bp, or the MORE VOLATILE compound) will be present to a greater extent in the vapor than in the liquid. THE VAPOR IS RICHER THAN THE LIQUID IN THE MORE VOLATILE COMPONENT.

PRACTICAL CONSEQUENCE OF THE EQUATION: If one of the P^oN terms is much greater than the other (usually because one bp is much lower than the other) the two components may be separated easily by simple distillation, If the bps are close, the liquids will DISTILL TOGETHER.

3. Fractional Distillation

If the first-formed vapor is condensed and removed (i.e., distilled), the new liquid which has been condensed is richer than the original liquid in the more volatile compound. Further distillation of this new liquid mixture will give (in the first drops) a distillate which is even richer in the more volatile compound. A series of separate distillations is impractical, but the separations can be achieved by using a FRACTIONATION COLUMN (fractional distilling column). Such a column works by allowing many condensations and vaporizations on its surfaces before the liquid is finally collected. The first liquid collected from the column will be mostly the more volatile compound; the second liquid will be the less volatile compound.

In a successful fractionation (separation) there is a sharp rise in temperature after the first component has distilled off, and the second component begins to distill. In a poor separation, there is a gradual rise in temperature throughout the distillation, since both components are contributing substantially to the total vapor pressure over the temperature range.

The graphs below show typical temperature vs. volume curves.

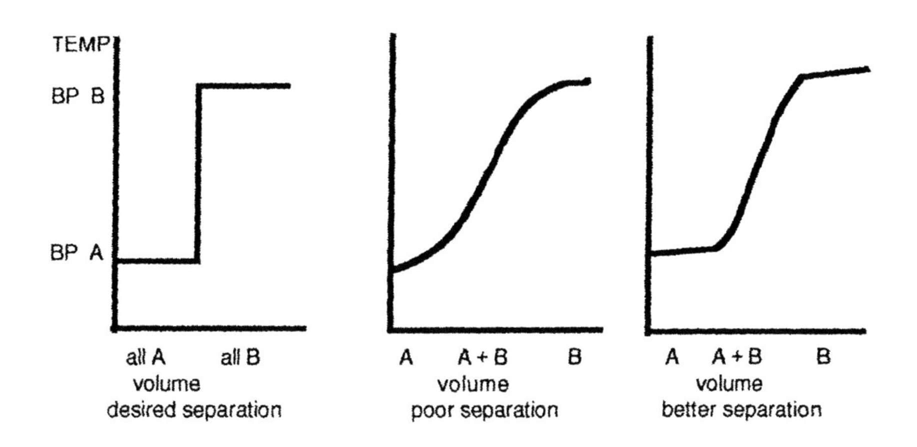

The fractional distillation apparatus we use (distillation column packed with copper sponge) can separate two liquids whose boiling points differ by as little as 20-30 °C. Liquids whose boiling points differ by more than 50 to 60 °C can be separated by a simple distillation.

Experiment 1: FRACTIONAL DISTILLATION OF A MIXTURE OF CYCLOHEXANE & TOLUENE

Cyclohexane	Toluene
MW: 84.16	MW: 92.14
density: 0.78 g/mL	density: 0.87 g/mL
mp: 6.5 ^{o}C bp: 81.4 ^{o}C	mp: -95 ^{o}C bp: 110.8 ^{o}C

Combine 15 mL of cyclohexane and 15 mL of toluene in a 50 mL round-bottomed flask, add one or two boiling chips to the flask to promote even boiling, and then assemble the fractional distillation apparatus shown in the Figure labeled "Fractional Distillation" in "Apparatus and Techniques". The fractionating column is packed with copper sponge (obtained from the stockroom).

Make sure that the fractionating column is not packed too tightly, so vapor and liquid can move freely through the column. Clamp the flask high enough above the bench top so that the heating mantle can easily be lowered away from the flask. Also make sure the electrical cord attached to the heating mantle will reach the rheostat-controlled outlet. Each ground joint is greased by putting one stripe of grease lengthwise around the male joint and pressing the joint firmly into the other with gentle twisting. The air is thus eliminated and the joint will appear almost transparent. (Do not use excess grease as it will contaminate the product.) Water enters the condenser at the tablature nearest the receiver (lowest tablature). Only a very small stream of water is needed; too much water pressure will cause the tubing to pop off. A Keck clamp is used to hold the condenser to the distillation head. Note that the bulb or tip of the thermometer is just below the opening into the side arm of the distillation head (see the Figure labeled "Fractional Distillation" in "Apparatus and Techniques"). The side arm of the distillation head is connected to the condenser which is connected to the vacuum adapter. Also use a Keck clamp to attach the vacuum adapter to the end of the condenser. The bottom of the vacuum adapter is positioned just above the top of a small (25 mL) graduated cylinder held by a clamp. Have your instructor or a TA check your apparatus before you proceed further.

Gradually turn up the heat to the electric heating mantle until the mixture of cyclohexane and toluene just begins to boil (a setting of 20 to 30 should work). Heat slowly at first. A ring of condensate will rise slowly through the column. The rise should be very gradual, in order that the column can acquire a uniform temperature gradient. Do not apply any more heat until you are sure that the ring of condensate has stopped rising; then increase the heat gradually. In a properly conducted operation, the vapor-condensate mixture reaches the top of the column only after several minutes. Once distillation has commenced, it should continue steadily without any drop in temperature at a rate not greater than 1 mL in 1.5-2 min (1 drop every few seconds). Observe the flow and keep it steady by slight increases in heat as required. Protect the column from drafts by wrapping a cloth around it. Record the temperature as each one-half milliliter of distillate collects.

When 7 mL of distillate has been collected, remove the graduated cylinder and collect the next 0.5 - 1 mL (approximate) of distillate in a small test tube (labeled "7 mL"), then replace the graduated cylinder. Repeat this with another test tube (labeled "20 mL") after 20 mL of distillate has been collected. Replace the graduated cylinder, and stop the distillation when 25 mL has been collected. Cork the test tubes and seal them each with Parafilm. Keep the test tubes for GC analysis next lab period, and pour the distillate and residue in the distillation flask into the ORGANIC WASTE container in the hood.

Use graph paper or a spreadsheeet program (EXCEL) to plot a distillation curve (volume collected as the x-axis, observed temperature as the y-axis). Be sure to use appropriate scales for each axis. Label the graph with a title, and label parts of the graph which correspond to the distillation of cyclohexane, toluene, and both compounds. Your instructor will tell you whether to paste the graph into your notebook or to turn it in with your report.

GC analysis. Your instructor will tell you how to prepare your samples from the distillates collected at 7 and 20 mL for GC analysis. Label each chromatogram with your name, date and sample name (e.g., 7 mL distillate), and label the peaks which correspond to the solvent, cyclohexane, and toluene. Using information provided by your instructor or from the table of "Gas Chromatography Conditions" at the back of this book, enter the exact GC conditions used in obtaining your chromatograms in your notebook. In your notebook (not on the chromatogram), also calculate the percentage of cyclohexane and toluene in each sample. Your instructor will tell you if the original copy of the chromatogram should be attached to your notebook or your written report.

USE OF THE AGILENT 7890A GAS CHROMATOGRAPHY SYSTEM

This instrument is equipped with a flame-ionization detector and a 5 m long wide-bore capillary column packed with HP-1 methyl siloxane gum. The carrier gas is helium. The gas chromatograph is very sensitive. Make sure your glassware is clean!

BE CAREFUL WITH THE MICROLITER SYRINGES!!

Sample Preparation for GC Analysis:

A. Liquid Samples:

1. Select a 50 mL beaker and label it as "GC/IR" with your name on it.

2. Label two clean small test tubes as "1" and "2". Place these two tubes in the above beaker.

3. Transfer about 1 mL of your final product (distillate or distillate fraction) into test tube "1".

4. Add about 0.40 to 0.5 mL dichloromethane (CH_2Cl_2) into test tube "2".

5. Choose a capillary syringe and rinse it with clean dichloromethane 3 to 4 times in order to ensure it is not contaminated.

6. Withdraw about 3 μL (micro liters) of the sample from test tube "1" using the capillary syringe.

7. Lower the volume of the syringe to around 1 μL by returning the excess volume into the test tube "1". Note: It is IMPORTANT not to have more than about 1 micro liter of volume in the syringe; larger volumes will overload the GC column and give unsatisfactory results.

8. Inject the remaining portion of the sample in the syringe (1 μL) into test tube "2" by immersing the tip of the needle into the solvent.

9. Gently shake/swirl test tube "2" to fully mix the materials.

10. Rinse the syringe with clean dichloromethane 3 to 4 times in order to clean it.

11. Remove 3 μL of the sample from the test tube "2" using a previously cleaned syringe (be careful not to introduce air bubbles into the syringe barrel when withdrawing the liquid).

12. Carefully, lower the volume of the syringe to 1 μL.

13. Check that the temperature has equilibrated and the GC is READY, inject this 1 μL of your sample into the GC, using the technique taught by your instructor , and press START.

14. Rinse the syringe with clean dichloromethane 3 to 4 times in order to clean it.

    If you have several samples to analyze (for example, two distillation fractions, or both starting material and product) analyze all samples on the same GC on the same day under the same conditions. Keep track of which GC instrument you use. (If there is anything strange about your results, run a blank as well (inject pure solvent only).)

15. Wait until the temperature program has finished and the integrator has printed out the sample analysis.

16. On the chart paper, record your name, the date, which GC you used (A, B, C, or D), your sample number, and the GC conditions (initial temperature and time, program rate, final temperature, etc (see the table of "Gas Chromatography Conditions" in the back of the book for the entire list of conditions). Also record the GC conditions in your notebook, and paste/staple the GC chart output in your notebook

17. If others are waiting, let them know the instrument is available.

18. After lab is over record all your calculations of the % composition of each of your samples in your notebook.

B. Solid Samples:

1. Select a 50 mL beaker and label it as "GC/IR" with your name on it.

2. Place a clean small test tube and a disposable pipette in the beaker.

3. Add about 0.5 mL of dichloromethane (CH_2Cl_2) into the above test tube using the disposable pipette .

4. Sample a **very** small amount of your solid using the tip of another clean and dry disposable pipette and place it into the test tube. Alternatively weigh out 0.020 g of the solid and add it to the dichloromethane in the test tube. Agitate gently to dissolve the solid.

5. Dissolve the solid in the solvent by repeatedly pulling the solution from the test tube back and forth into the disposable pipette.

6. Remove 3 μL of the sample from the test tube "2" using a previously cleaned syringe (be careful not to introduce air bubbles into the syringe barrel when withdrawing the liquid).

7. Carefully lower the volume of the syringe to 1 μL.

8. Check that the temperature has equilibrated and the GC is READY, inject this 1 μL of your sample into the GC, using the technique taught by your instructor , and press START.

9. Rinse the syringe with clean dichloromethane 3 to 4 times in order to clean it.

   If you have several samples to analyze (for example, two distillation fractions, or both starting material and product) analyze all samples on the same GC on the same day under the same conditions. Keep track of which GC instrument you use. (If there is anything strange about your results, run a blank as well (inject pure solvent only).)

10. Wait until the temperature program has finished and the integrator has printed out the sample analysis.

11. On the chart paper, record your name, the date, which GC you used (A, B, C, or D), your sample number, and the GC conditions (initial temperature and time, program rate, final temperature, etc (see the table of "Gas Chromatography Conditions" in the back of the book for the entire list of conditions). Also record the GC conditions in your notebook, and paste/staple the GC chart output in your notebook

12. If others are waiting, let them know the instrument is available.

13. After lab is over record all your calculations of the % composition of each of your samples in your notebook.

Typical GC results for the distillation experiment are:

Compound	Retention Time (min)
solvent (CH_2Cl_2)	0.04
cyclohexane	0.62
toluene	1.18

SEPARATION OF ACIDIC AND NEUTRAL COMPOUNDS BY EXTRACTION

As indicated by their pK_a values, benzoic acid ($pK_a = 4.2$) is much more acidic than 2-naphthol ($pK_a = 9.5$). Accordingly benzoic acid will react with sodium bicarbonate, a weak base, but 2-naphthol will not.

$$C_6H_5COOH + NaHCO_3 \longrightarrow [C_6H_5COO^-][Na^+] + CO_2 + H_2O$$

$$\text{2-naphthol} + NaHCO_3 \longrightarrow \text{NR}$$

If a solution containing benzoic acid, 2-naphthol, and 1,4-dimethoxybenzene dissolved in an organic solvent is reacted with aqueous sodium bicarbonate, benzoic acid will be converted to the water soluble benzoate anion while 2-naphthol and 1,4-dimethoxybenzene will remain in the organic layer since they do not react with sodium bicarbonate. After the benzoic acid is removed from the organic layer, 2-naphthol can be converted to the water soluble 2-naphtholate anion by reaction with aqueous sodium hydroxide, a strong base.

$$\text{2-naphthol (OH)} + NaOH \longrightarrow [\text{2-naphtholate } (O^-)][Na^+] + H_2O$$

Thus after both the sodium bicarbonate and then the sodium hydroxide extractions only 1,4-dimethoxybenzene remains in the organic layer. 1,4-Dimethoxybenzene does not react with either sodium bicarbonate or sodium hydroxide.

$$\text{1,4-dimethoxybenzene} \xrightarrow{NaHCO_3} NR$$

$$\text{1,4-dimethoxybenzene} \xrightarrow{NaOH} NR$$

Neutralization of an aqueous solution of sodium benzoate with HCl causes benzoic acid to precipitate. Similarly, neutralization of an aqueous solution of sodium 2-naphtholate with HCl causes 2-naphthol to precipitate. If the aqueous solution of

$$[C_6H_5COO^-][Na^+] + HCl \longrightarrow C_6H_5COOH + [Na^+][Cl^-]$$

$$[C_{10}H_7O^-][Na^+] + HCl \longrightarrow C_{10}H_7OH + [Na^+][Cl^-]$$

sodium benzoate also contains sodium bicarbonate, the sodium bicarbonate will also react with HCl to produce carbon dioxide gas.

$$NaHCO_3 + HCl \rightarrow CO_2\uparrow + NaCl$$

Experiment 2: SEPARATION OF ACIDIC AND NEUTRAL SUBSTANCES

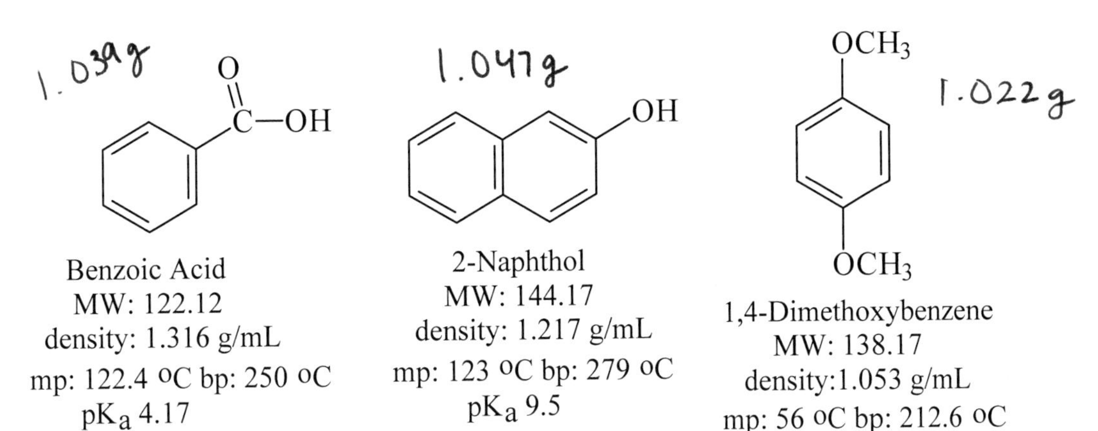

Weigh out 1.0 g each of benzoic acid, 2-naphthol, and 1,4-dimethoxybenzene. Dissolve the solids in a total of 30 mL of dichloromethane (CH_2Cl_2) in an Erlenmeyer flask. After the solids have dissolved, transfer the mixture to a 125-mL separatory funnel. The flask may be rinsed with about 2mLof dichloromethane, which is also added to the separatory funnel. Add 10 mL of water to the separatory funnel followed by 20 mL of a 5% aqueous solution of sodium bicarbonate. Stopper the funnel and gently swirl the contents (see "Use of a Separatory Funnel" in "Apparatus and Techniques"). Vent the carbon dioxide gas that is liberated, and then shake the mixture thoroughly with frequent venting of the funnel. Allow the layers to separate completely and then draw off the organic (lower) layer into a 100 mL beaker. Be sure to remove the stopper from the separatory funnel before you try to draw off anything. Pour the aqueous (upper) layer into a 125-mL Erlenmeyer flask or a beaker (labeled "Carboxylic Acid"). Return the organic layer to the separatory funnel, add 20 mL of 5% aqueous sodium hydroxide to the separatory funnel, and then shake the mixture thoroughly, allow the layers to separate. Draw off the organic (lower) layer into a clean 100 mL beaker. Pour the aqueous (upper) layer into another 125-mL Erlenmeyer flask or beaker (labeled "Phenol"). Return the organic layer to the separatory funnel, add 5 mL of water to the separatory funnel, shake the mixture thoroughly, and separate the layers as before, adding the aqueous layer to the beaker or flask labeled "Phenol". Return the organic layer to the separatory funnel, add 15 mL of a saturated aqueous solution of sodium chloride to the separatory funnel, shake the mixture thoroughly, and allow the layers to separate. Draw off the organic (lower) layer into a clean 125-mL Erlenmeyer flask (labeled "1,4-DMB"), not a beaker. Add about 4 g of anhydrous sodium sulfate to the organic layer, swirl the mixture, and set it aside. Discard the aqueous layer (see "Before you leave" below).

Acidify the contents of the flask or beaker labeled "Phenol" by dropwise addition of 6M hydrochloric acid with a Pasteur pipette. Use litmus paper to determine when the

4.023g Anhydrous solution

solution is acidic (an acidic solution will turn blue litmus paper red). Use a clean stirring rod to transfer a drop of the solution to the litmus paper. When the solution is acidic, cool the flask in an ice bath. You should have a voluminous precipitate.

Cautiously add 6M hydrochloric acid dropwise to the contents of the flask or beaker labeled "Carboxylic Acid" with a Pasteur pipette. Bubbling or foaming will occur due to formation of CO_2 gas from the reaction of sodium bicarbonate with the acid. Again use litmus paper to determine when the solution is acidic by transferring a drop of the solution to the litmus paper with a clean stirring rod. When the solution is acidic cool the flask in an ice bath. You should have a voluminous precipitate.

Decant (pour off) the dichloromethane from the flask labeled "1,4-DMB" into a clean, tared (previously weighed) 100 mL beaker making sure you leave all of the drying agent behind. Rinse the drying agent with about 5 mL additional dichloromethane to ensure complete transfer of the product and add the rinse dichloromethane to the dichloromethane solution already in the tared beaker. Simply cover the beaker (no boiling chip necessary) with a Kimwipe held in place by a rubber band, label the beaker with your name and section number, and place the covered beaker in your drawer until the next laboratory period so that the dichloromethane can evaporate slowly. Alternatively, put a boiling chip in the beaker and evaporate the dichloromethane on a hot plate in the hood. After all the dichloromethane has evaporated determine the weight of the 1,4-dimethoxybenzene. 1,4-Dimethoxybenzene can be recrystallized from methanol (use 10 mL of methanol per gram of crude 1,4-dimethoxybenzene).

Isolate the precipitated 2-naphthol from the flask or beaker labeled "Phenol" by vacuum filtration using a Büchner funnel (see "Vacuum Filtration" in "Apparatus and Techniques"), wash it on the filter with about 5 mL of ice cold water, and then allow it to partially air dry. Carefully remove the filter paper disk containing the 2-naphthol from the Büchner funnel and place it in a plastic weighing boat. Cover the weighing boat with a loose Kimwipe and place the covered weighing boat in your drawer. The solid will be completely dry by the next laboratory period. Determine the weight of the 2-naphthol. Follow the same procedure to isolate, wash, and dry the benzoic acid from the flask or beaker labeled "Carboxylic Acid". Benzoic acid may be recrystallized from boiling water (use 20 mL of water per gram of crude benzoic acid). The solubility of benzoic acid in water is 1.9 g/L at 0 °C and 68 g/L at 95 °C.

Determine the melting point (see "Melting Points" in "Apparatus & Techniques") and weight (amount recovered) of each of the three compounds. Calculate the percent recovery of each compound. Analyze the benzoic acid, 2-naphthol, and 1,4-dimethoxybenzene samples you isolated by Gas Chromatography (GC, see "Gas Chromatography" in "Apparatus & Techniques"). Your instructor will tell you how to make up the sample solutions. Analyze each of the compounds you isolated by Thin Layer Chromatography (TLC, see "Thin Layer Chromatography" in "Apparatus & Techniques").

To analyze your samples by TLC, prepare a developing chamber by filling a clean dry 100 ml beaker to about 4 mm high with the developing solvent (a mixture of methanol and dichloromethane (5: 95, i.e. 5% methanol : 95% dichloromethane)). Insert a piece of filter paper in the chamber with one of its ends submerged in the solvent so as to saturate the chamber with the solvent vapor. Then cover the beaker tightly with aluminum foil.

Prepare a solution of each solid sample (~1 mg) in a few drops of dichloromethane or fill the narrow end of a Pasteur pipette to a depth of about 1.0 cm with your solid sample. Dissolve this amount of solid in 8-10 drops of dichloromethane in a small test tube. Make a pencil mark about 6 mm from the bottom of the TLC plate (do NOT use a pen). Using a fine capillary tube or pipette tip, spot each sample solution as close to this pencil line as possible. (Be careful not to make a large spot and make sure that each spot is above the solvent line once you insert the plate in the developing chamber). Make a second pencil mark about 3-4 mm from the top of the plate. The stationary phase for this experiment is silica gel ($SiO_2 \bullet xH_2O$) coated on a thin sheet of aluminum.

Develop your plate in the developing chamber, so that the solvent travels up the plate by capillary action carrying components of the mixture along the silica gel stationary phase. Remove the plate when the solvent just reaches the second pencil line that is about 3-4 mm from the top of the plate. This line is the position of your solvent front. Wait about 30-40 seconds to allow the solvent to completely evaporate from the plate, and then observe the plate under a UV lamp (Do NOT look directly at the UV light) . Mark the outline of each spot with a pencil and calculate the R_f value for each observed spot.

For this experiment, you will run three TLC plates with three solutions spotted on each plate.

Plate 1: Spot your isolated benzoic acid along with separate authentic samples of benzoic acid and 2-naphthol.

Plate 2: Spot your isolated 2-naphthol along with separate authentic samples of 2-naphthol and benzoic acid.

Plate 3: Spot your isolated 1,4-dimethoxybenzene along with separate authentic samples of 1,4-dimethoxybenzene and 2-naphthol

Make a note of the number of spots in each of your isolated products after developing the plate and identify each spot. Comment on the relative polarity of the compounds as seen on the TLC plate. Make a sketch of each of your developed TLC plates in your laboratory notebook that shows the relative positions of each spot seen under the UV lamp. Also attach each TLC plate with the spots outlined in pencil in your notebook.

Hand in each compound in a neatly labeled test tube. The label should have the following information: your name, your ID#, the compound name, the weight of the compound, and the observed melting point or melting range of the compound.

The percent recovery of compound X is calculated as follows:

$$\% \ recovery \ of \ X = \frac{Final \ weight \ of \ X}{Initial \ weight \ of \ X} \times 100$$

Before You Leave: Combine all aqueous layers, washes, and filtrates. Dilute with water, neutralize using either sodium carbonate or dilute hydrochloric acid and then discard with other aqueous wastes. The methanol filtrate, any dichloromethane, and remaining developing solvent go in the organic waste container. Allow the dichloromethane to evaporate from the sodium sulfate in the hood. Then the sodium sulfate can be placed in the nonhazardous solid waste container.

NAME Jarius Garner I.D. @02877197

RECOVERY REPORT EXPERIMENT 2 SEPARATION OF ACIDIC AND NEUTRAL SUBSTANCES BY EXTRACTION

BENZOIC ACID

WEIGHT ORIGINAL SAMPLE 1.039 g $R_f = \frac{1.5 \text{ in}}{2 \text{ in}} = 0.75$

WEIGHT, RECOVERED SAMPLE 1.644 g

PERCENT RECOVERY 158 %

MELTING POINT (OBSERVED RANGE) 115 °C (122.4°C)

2-NAPHTHOL

WEIGHT ORIGINAL SAMPLE 1.047 g $R_f = \frac{1.25}{2} = 0.63$

WEIGHT, RECOVERED SAMPLE 0.033 g

PERCENT RECOVERY 3 %

MELTING POINT (OBSERVED RANGE) 118 °C (123°C)

1,4-DIMETHOXYBENZENE

WEIGHT ORIGINAL SAMPLE 1.022 g $R_f = \frac{1}{2} = 0.50$

WEIGHT, RECOVERED SAMPLE 0.048 g

PERCENT RECOVERY 4.7 %

MELTING POINT (OBSERVED RANGE) 65 °C (56°C)

THIS PAGE SHOULD BE TURNED IN WITH YOUR SAMPLES. THE REQUIRED INFORMATION SHOULD ALSO BE RECORDED IN YOUR NOTEBOOK. THE LABEL ON EACH SAMPLE MUST INCLUDE YOUR NAME AND THE NAME OF THE COMPOUND.

TLC

Stationary → Solu. = 2 inches = 50.8 mm

CA = 1 1/2 in = 38.1 mm

2-N = 1 1/4 in = 31.2 mm

DMB = 1 in = 25.4 mm

$$\text{cyclohexanol} + H_3PO_4 \rightarrow \text{protonated cyclohexanol } (-O^{+}H_2) + H_2PO_4^{-}$$

$$\rightarrow \text{cyclohexyl carbocation} + H_2O$$

$$\rightarrow \text{cyclohexene} + H_2O$$

ALKENES FROM ALCOHOLS: CYCLOHEXENE FROM CYCLOHEXANOL

Many alkenes can be prepared from alcohols by an acid catalyzed dehydration reaction. When a secondary alcohol, such as cyclohexanol, is heated with a concentrated

$$\text{cyclohexanol} + H_3PO_4 \rightleftharpoons \text{protonated cyclohexanol} + H_2PO_4^-$$

$$\text{protonated cyclohexanol} \rightleftharpoons \text{carbocation} + H_2O \rightleftharpoons \text{cyclohexene} + H_3O^+$$

acid, such as phosphoric acid, the protonated alcohol can lose water to generate a carbocation. Loss of hydrogen ion by the carbocation will give the corresponding alkene.

Since the boiling points of cyclohexene (83 °C) and cyclohexanol (161 °C) are so different, the cyclohexene product can be separated from the starting materials by simple distillation. Because each of the above reactions is reversible, removing the product as it forms drives the reaction towards product formation. In order to recover all the cyclohexene an additional solvent called the chaser solvent, toluene (bp 111 °C) may be added after most of the cyclohexene has been recovered. Since the boiling point of toluene is also higher than that of cyclohexene, the distillation can be continued until all of the cyclohexene has been recovered. Some toluene will also be found in the distillate. Ultimately, pure cyclohexene is obtained by fractional distillation of the cyclohexene/toluene mixture. Your instructor will tell you whether or not to use the toluene chaser.

Experiment 3: CYCLOHEXENE FROM THE DEHYDRATION OF CYCLOHEXANOL

OH

H_3PO_4

\+ H_2O

Cyclohexanol	Cyclohexene
MW: 100.16	MW: 82.14
density: 0.96 g/mL	density: 0.81 g/mL
mp: 25 °C bp: 161 °C	mp: -103.7 °C bp: 83.3 °C

Transfer 10.0 g of cyclohexanol and then 5 mL of concentrated (85%) phosphoric acid to a 50 mL round-bottomed flask, then swirl the flask to mix the layers. Some heat may be evolved. Add one or two boiling chips to the flask, and then assemble the apparatus for fractional distillation (see the Figure labeled "Fractional Distillation" in "Apparatus and Techniques") that has the vacuum adapter positioned directly into an ice-cooled test tube in a 125 mL Erlenmeyer flask. No column packing is necessary.

If your instructor tells you to use a toluene chaser, follow the procedure in paragraph A, below. Otherwise, follow the procedure in paragraph B below.

A. Distill the mixture until the residue in the flask has a volume of approximately 5 mL and very little distillate is being formed, noting the temperature range over which you collect distillate. Then let the entire apparatus cool a little, remove the thermometer, and quickly pour 10 mL of toluene (the chaser solvent) into the top of the column through a long-stemmed funnel. Replace the thermometer. Note the amount of liquid in the boiling flask and continue distilling again until the volume of the volume in the flask has been reduced by about half. Remove the heating mantle and allow the apparatus to cool. Pour the contents of the test tube (distillate) into a small separatory funnel. Add an equal volume of saturated sodium chloride solution, shake the funnel vigorously, and remove the lower (aqueous) layer. transfer the upper (organic) layer into a clean, dry flask, and add 5 g of anhydrous sodium sulfate to dry the organic layer (see "Drying Agents" in "Apparatus and Techniques"). Decant the liquid into a clean and **dry** 50 mL round bottom flask, and set up the apparatus for a fractional distillation (see the Figure labeled "Fractional Distillation" in "Apparatus and Techniques"). Be sure you rinsed the condenser and vacuum adapter out with acetone in order to remove any water left over from the initial distillation. Don't forget to add one or two boiling chips to the distillation flask. The fractionating column should be packed with copper mesh. Collect the material distilling around 83 °C into a tared, ice-cooled 18x150 mm test tube supported in a 125 mL Erlenmeyer flask. If the cyclohexene was dried completely, it will be clear and colorless. If it was not dried completely, it will be cloudy. Around 5 g of cyclohexene will be obtained (5 g is not the theoretical yield). Record your yield in grams, and calculate your percent yield in your notebook. Save the cyclohexene in a tightly stoppered, Parafilm wrapped test tube for later GC and IR analysis. After you have performed the GC and IR analyses, record the results and any appropriate calculations or interpretations in your notebook.

B. Distill the mixture until the residue in the flask has a volume of approximately 5 mL and very little distillate is being formed, noting the temperature range over which you collect distillate. Remove the heating mantle and allow the apparatus to cool. Pour the contents of the test tube (distillate) into a small separatory funnel. Add an equal volume of saturated sodium chloride solution, shake the funnel vigorously, and remove the lower (aqueous) layer. Transfer the organic (upper) layer into a clean, dry flask, and add 5 g of anhydrous sodium sulfate to remove water from the organic layer (see "Drying Agents" in "Apparatus and Techniques"). Decant the liquid into a clean and **dry** 50 mL round bottom flask, and set up the apparatus for a simple distillation (see the Figure labeled "Simple Distillation" in "Apparatus and Techniques"). Be sure you rinsed the condenser and vacuum adapter out with acetone in order to remove any water left over from the initial distillation. Don't forget to add one or two boiling chips to the distillation flask. Collect the material distilling around 83 °C into tared, ice-cooled 18x150 mm test tube supported in a 125 mL Erlenmeyer flask. If the cyclohexene was dried completely, it will be clear and colorless; if it is not completely dry, it will be cloudy. Around 5 g of cyclohexene will be obtained (5 g is not the theoretical yield). Record your yield in grams, and calculate your percent yield in your notebook. Save the cyclohexene in a tightly stoppered, Parafilm-wrapped test tube for later GC and IR analysis. After you have performed the GC and IR analyses, record the results and any appropriate calculations or interpretations in your notebook.

Before You Leave: The aqueous solutions (washes) should be diluted with water and neutralized. The neutralized solution can be flushed down the drain with a large excess of water. All the distillation pot residues should be placed in the waste solvents container. Sodium sulfate that is free of solvent (toluene) can be placed in the nonhazardous solid waste container or dissolved in water and flushed down the drain followed by lots of water.

The IR spectra of cyclohexanol and cyclohexene are shown below:

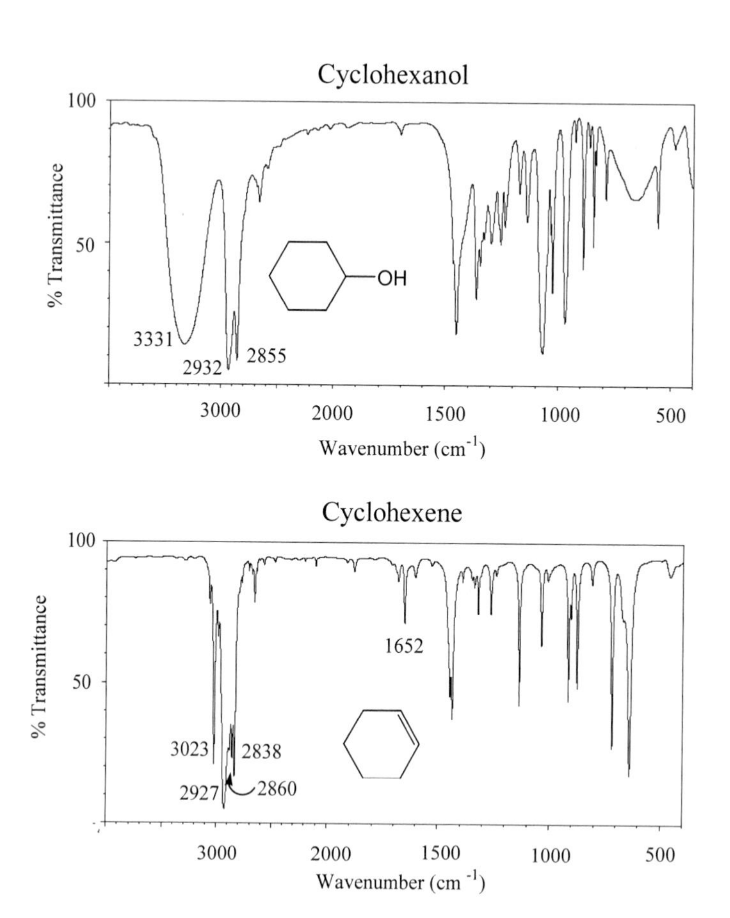

NAME Jarius Garner I.D. C02877197

YIELD REPORT EXPERIMENT 3: SYNTHESIS OF CYCLOHEXENE FROM THE DEHYDRATION OF CYCLOHEXANOL

WEIGHT OF CYCLOHEXANOL (Menthol)	10.0	g
MOLES OF CYCLOHEXANOL	0.064	mol
THEORETICAL YIELD OF CYCLOHEXENE		mol
THEORETICAL YIELD OF CYCLOHEXENE	8.848	g
ACTUAL YIELD OF CYCLOHEXENE	1.74	g
PERCENT CYCLOHEXENE IN DISTILLATE		%
PERCENT YIELD OF CYCLOHEXENE	19.7	%
BOILING POINT (RANGE) OF PRODUCT	95 - 130	°C

$\% = \frac{1.74}{8.848} \times 100$

$= 19.7\%$

THIS PAGE SHOULD BE TURNED IN WITH YOUR SAMPLE. THE REQUIRED INFORMATION SHOULD ALSO BE RECORDED IN YOUR NOTEBOOK. THE LABEL ON EACH SAMPLE MUST INCLUDE YOUR NAME, THE NAME OF THE COMPOUND, THE WEIGHT OF THE COMPOUND, AND THE BOILING RANGE OF THE COMPOUND.

YIELD CALCULATIONS

The success of a synthesis is measured by the quantity and purity of the product. The quantity (weight) of product obtained is compared to the amount which is expected from the balanced equation and molecular weights. The expected weight of the product is called the theoretical yield. The actual weight of the product (actual yield) is compared to the theoretical yield and expressed as a percentage.

$$\textit{percent yield} = 100 \cdot \left(\frac{\textit{actual yield}}{\textit{theoretical yield}} \right) = xx\ \%$$

Example: If 6.23 g of product was actually obtained and the theoretical yield was 8.67 g, the percent yield will be:

$$\textit{percent yield} = 100 \cdot \left(\frac{6.23}{8.67} \right) = 71.9\ \%$$

The theoretical yield is based on the balanced equation and the actual weight of the reagents used. For a general equation:

$$a\,A + b\,B \rightarrow x\,X + y\,Y$$

The theoretical yield in grams of product X can be calculated from the original weight of A in grams and the molecular weight of A in grams/mole:

$$\textit{theoretical yield } X = (\textit{wt. } A) \cdot \left(\frac{1 \textit{ mole } A}{\textit{mol. wt } A} \right) \cdot \left(\frac{x \textit{ mole } X}{a \textit{ mole } A} \right) \cdot \left(\frac{\textit{mol. wt. } X}{1 \textit{ mole } X} \right) = \textit{wt. } X$$

and from the original weight of B in grams and the molecular weight of B in grams/mole:

$$\textit{theoretical yield } X = (\textit{wt. } B) \cdot \left(\frac{1 \textit{ mole } B}{\textit{mol. wt. } B} \right) \cdot \left(\frac{x \textit{ mole } X}{b \textit{ mole } B} \right) \cdot \left(\frac{\textit{mol. wt. } X}{1 \textit{ mole } X} \right) = \textit{wt. } X$$

The theoretical yield will be the smaller calculated value for the weight of X.

In the following examples the molecular weights are as follows: A, 100 g/mole;

B, 150 g/mole; and X, 200 g/mole.

Example 1: For the balanced equation A + B → X + Y
starting with 10.0 g of A, and 20.0 g of B; what is the theoretical yield of X?

Solution: Based on the weight of A the theoretical yield of X is:

$$\textit{theoretical yield } X = (10.0\ g\ A)\cdot\left(\frac{1\ mole\ A}{100g\ A}\right)\cdot\left(\frac{1\ mole\ X}{1\ mole\ A}\right)\cdot\left(\frac{200g\ X}{1\ mole\ X}\right) = 20.0\ g\ X$$

Based on the weight of B:

$$\textit{theoretical yield } X = (20.0\ g\ B)\cdot\left(\frac{1\ mole\ B}{150\ g\ B}\right)\cdot\left(\frac{1\ mole\ X}{1\ mole\ B}\right)\cdot\left(\frac{200\ g\ X}{1\ mole\ X}\right) = 26.7\ g\ X$$

Since 20.0 is smaller than 26.7, the theoretical yield of X is 20.0 g, and A is the limiting reagent; B is in excess, and increasing the amount of B will not change the theoretical yield.

Example 2: For the balanced equation A + 2 B → X + Y
starting with 10.0 g of A, 20.0 g of B; what is the theoretical yield of X?

Solution: Based on the weight of A:

$$\textit{theoretical yield } X = (10.0\ g\ A)\cdot\left(\frac{1\ mole\ A}{100\ g\ A}\right)\cdot\left(\frac{1\ mole\ X}{1\ mole\ A}\right)\cdot\left(\frac{200\ g\ X}{1\ mole\ X}\right) = 20.0\ g\ X$$

Based on the weight of B:

$$\textit{theoretical yield } X = (20.0\ g\ B)\cdot\left(\frac{1\ mole\ B}{150\ g\ B}\right)\cdot\left(\frac{1\ mole\ X}{2\ mole\ B}\right)\cdot\left(\frac{200\ g\ X}{1\ mole\ X}\right) = 13.3\ g\ X$$

In this case B is the limiting reagent and the theoretical yield of X is 13.3 g.

Example 3: For the balanced equation $3\,A + 2\,B \rightarrow 2\,X + Y$
starting with 10.0 g of A, 20.0 g of B; what is the theoretical yield of X?

Solution: Based on the weight of A:

$$\textit{theoretical yield } X = (10.0\ g\ A)\cdot\left(\frac{1\ mole\ A}{100\ g\ A}\right)\cdot\left(\frac{2\ mole\ X}{3\ mole\ A}\right)\cdot\left(\frac{200\ g\ X}{1\ mole\ X}\right) = 13.3\ g\ X$$

Based on the weight of B:

$$\textit{theoretical yield } X = (20.0\ g\ B)\cdot\left(\frac{1\ mole\ B}{150\ g\ B}\right)\cdot\left(\frac{2\ mole\ X}{2\ mole\ B}\right)\cdot\left(\frac{200\ g\ X}{1\ mole\ X}\right) = 26.7\ g\ X$$

Here, A is the limiting reagent and the theoretical yield of X is 13.3 g.

YIELD PROBLEMS

Where excess reagents are not specified in the problem, it is necessary to use the balanced equation to determine the limiting reagent, in order to calculate the theoretical yield. The calculation of the theoretical yield is the first procedure in any case.

ATOMIC WEIGHTS: H = 1 C=12 O = 16 N = 14 S = 32 Cl = 35.5 Br = 80

1. Cyclohexanone ($C_6H_{10}O$) is reduced with excess hydrogen under pressure to give cyclohexanol ($C_6H_{12}O$). In one experiment, 49 g of cyclohexanone was reduced and 40 g of cyclohexanol was isolated. Calculate the percent yield of cyclohexanol.

2. Benzene (50 g) was nitrated with an excess of nitric acid to give 55 g of nitrobenzene. Calculate the percent yield.

HNO_3 → NO_2

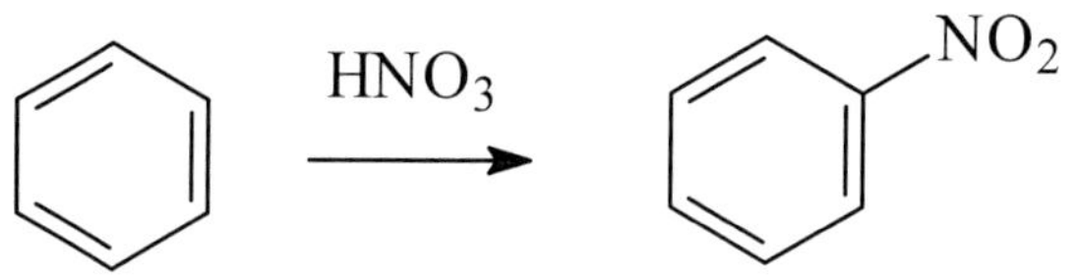

3. In the bromination of acetanilide, 8 g of bromine was used with 8 g of acetanilide (structure below) and 8 g of *p*-bromoacetanilide was isolated. What is the limiting reagent? Calculate the percent yield.

NH–C(=O)–CH_3 + Br_2 → Br, NH–C(=O)–CH_3 + HBr

acetanilide

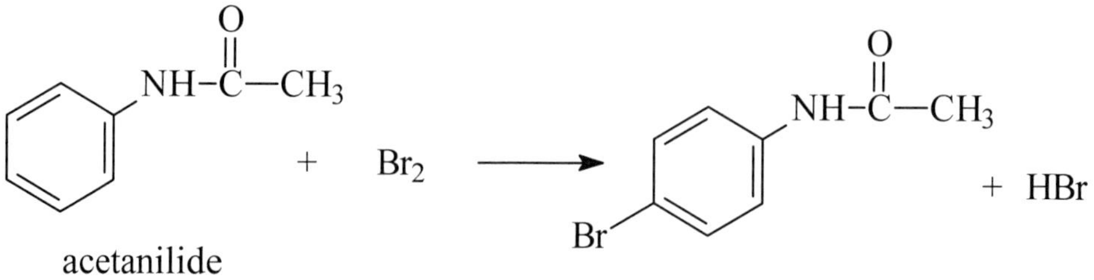

4. One of the reactions in the synthesis of a sulfa drug is shown below. In one synthesis, 18 g of acetanilide was allowed to react with 25 g of chlorosulfonic acid, and 15 g of the product shown was isolated. What is the limiting reagent? Calculate the percent yield.

$$C_6H_5\text{-}NH\text{-}C(=O)\text{-}CH_3 \xrightarrow{ClSO_3H} ClSO_2\text{-}C_6H_4\text{-}NH\text{-}C(=O)\text{-}CH_3 + HCl + H_2SO_4$$

acetanilide

OXIDATION OF ALCOHOLS

An organic compound which is **oxidized** loses electrons, and gains oxygen or loses hydrogen in its molecular formula.

An organic compound which is **reduced** gains electrons, and loses oxygen or gains hydrogen in its molecular formula.

Several important functional groups are related in oxidation series, as shown in the table below:

OXIDATION →		**← REDUCTION**
RCH_2-OH primary alcohol	R-CH=O aldehyde	R-CO_2H carboxylic acid
R_2CH-OH secondary alcohol	R_2C=O ketone	----- no oxidation product
R_3C-OH tertiary alcohol	----- no oxidation product	

Primary and secondary alcohols are oxidized by inorganic reagents to give aldehydes and ketones, respectively, but tertiary alcohols are inert. However, aldehydes are easily oxidized to form carboxylic acids, so special techniques or reagents must be used to prevent the further oxidation of the aldehyde product of the oxidation of a primary alcohol.

In general, secondary alcohols may be converted readily to ketones using a variety of inorganic oxidizing agents, such as permanganate, hypochlorite, or chromate. In this experiment the oxidizing reagent is chromic acid, H_2CrO_4. The chromium in chromic acid is in the +6 oxidation state, and three electrons are gained by the chromium to form the +3 oxidation state.

Chromic acid is not stable over time, and it is prepared by mixing a dichromate salt, usually sodium or potassium dichromate, with sulfuric acid:

$$Na_2Cr_2O_7 + 2\ H_2SO_4 \rightarrow H_2Cr_2O_7 + 2\ NaHSO_4$$

$$H_2Cr_2O_7 + H_2O \leftrightharpoons 2\ H_2CrO_4$$

The reaction of chromic acid with a secondary alcohol involves several steps, in which the oxidation state of chromium is changed from +6 to +3, but the slowest (rate-limiting) step involves the decomposition of a chromate ester of the alcohol to give the ketone:

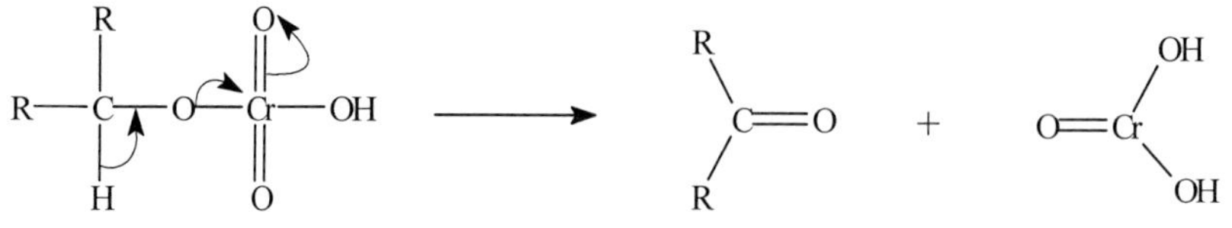

chromate ester — ketone + H_2CrO_3

In the chromium product of this step, H_2CrO_3, the chromium is in the +4 oxidation state. A further series of rapid steps ultimately brings chromium to the final +3 oxidation state. The progress of the reaction may be followed by observing the color change: chromic acid solution is yellow-orange, and the final chromium product, $Cr_2(SO_4)_3$, is dark blue-green.

The overall balanced equation for the oxidation reaction must be determined, as shown in the next section.

In the experiment, cyclohexanol is the secondary alcohol which is oxidized to cyclohexanone. After the reaction takes place, cyclohexanone must be isolated from the reaction mixture which also contains chromium salts and unreacted sulfuric acid. Adding water and an organic solvent and using extraction to isolate the cyclohexanone may be used, but an alternative method which gives the product and avoids handling the chromium salts is **steam distillation**. Steam distillation is used when a liquid compound has a high boiling point, and is not miscible with water. A large volume of water is added to the compound and the mixture is distilled. The total vapor pressure is equal to the sum of the vapor pressure of water and the vapor pressure of the organic compound. Near 100°, the vapor pressure of water is large (near 1 atm), and the vapor pressure of the organic compound is very small. However, the vapor pressure of the mixture will be 1 atm, and the mixture will boil. The condensate will contain mostly water, but the organic compound will be present also. The distillation is allowed to continue until all of the organic compound has distilled, and then the organic compound is separated from the water in a separatory funnel.

Steam distillation: $P_{total} = P_{water} + P_{organic}$

$1 \text{ atm} = \text{large} + \text{small}$

The organic compound and water are immiscible.

BALANCING OXIDATION-REDUCTION REACTIONS

In order to calculate yields, a balanced equation is required. There are several ways to balance oxidation-reduction reactions; the following is one method that works well.

Carry out the following steps for each half-reaction:

1. Balance the metal (or atom being oxidized/reduced) or carbon atoms.
2. Balance oxygen by adding H_2O.
3. Balance hydrogen by adding H^+.
4. Balance charge by adding electrons (e^-).

Combine the two half-reactions, multiplying so that the number of electrons is the same on each side of the equation. For reactions carried out under basic conditions, neutralize H^+ by adding hydroxide (OH^-) to each side of the equation. Add appropriate counter ions and salts as necessary.

Example 1: Potassium dichromate oxidation of a secondary alcohol.

$$R_2CH\text{-}OH + K_2Cr_2O_7 \rightarrow R_2C{=}O + Cr_2(SO_4)_3$$

For chromium half-reaction:

1. Balance Cr	$Cr_2O_7^{-2}$	→	$2\ Cr^{+3}$
2. Balance O	$Cr_2O_7^{-2}$	→	$2\ Cr^{+3} + 7\ H_2O$
3. Balance H	$Cr_2O_7^{-2} + 14\ H^+$	→	$2\ Cr^{+3} + 7\ H_2O$
4. Balance charge	$Cr_2O_7^{-2} + 14\ H^+ + 6\ e^-$	→	$2\ Cr^{+3} + 7\ H_2O$

For organic half-reaction:

1. Balance C	$R_2CH\text{-}OH$	→	$R_2C{=}O$
2. Balance O	$R_2CH\text{-}OH$	→	$R_2C{=}O$
3. Balance H	$R_2CH\text{-}OH$	→	$R_2C{=}O + 2\ H^+$

4. Balance charge $R_2CH\text{-}OH \rightarrow R_2C{=}O + 2\,H^+ + 2\,e^-$

Combine the two half-reactions. In this case the organic half-reaction must be multiplied by 3 to give 6 e^- in each.

$$Cr_2O_7^{-2} + 14\,H^+ + 6\,e^- \rightarrow 2\,Cr^{+3} + 7\,H_2O$$

$$3\,R_2CH\text{-}OH \rightarrow 3\,R_2C{=}O + 6\,H^+ + 6\,e^-$$

Simplify:

$$3\,R_2CH\text{-}OH + Cr_2O_7^{-2} + 8\,H^+ \rightarrow 3\,R_2C{=}O + 2\,Cr^{+3} + 7\,H_2O$$

Add counter ions (potassium (K^+) and sulfate (SO_4^{2-}) ions):

$$3\,R_2CH\text{-}OH + K_2Cr_2O_7 + 4\,H_2SO_4 \rightarrow 3\,R_2C{=}O + Cr_2(SO_4)_3 + 7\,H_2O + K_2SO_4$$

Example 2: Basic permanganate oxidation of a primary alcohol.

$$R\text{-}CH_2OH + KMnO_4 \rightarrow R\text{-}CO_2H + MnO_2$$

For manganese half-reaction:

1. Balance Mn $MnO_4^- \rightarrow MnO_2$

2. Balance O $MnO_4^- \rightarrow MnO_2 + 2\,H_2O$

3. Balance H $MnO_4^- + 4\,H^+ \rightarrow MnO_2 + 2\,H_2O$

4. Balance charge $MnO_4^- + 4\,H^+ + 3\,e^- \rightarrow MnO_2 + 2\,H_2O$

For organic half-reaction:

1. Balance C $R\text{-}CH_2OH \rightarrow R\text{-}CO_2H$

2. Balance O $R\text{-}CH_2OH + H_2O \rightarrow R\text{-}CO_2H$

3. Balance H $R\text{-}CH_2OH + H_2O \rightarrow R\text{-}CO_2H + 4 H^+$

4. Balance charge $R\text{-}CH_2OH + H_2O \rightarrow R\text{-}CO_2H + 4 H^+ + 4 e^-$

Combine half-reactions. In this case, in order to get 12 e^- in each reaction, the inorganic half-reaction must be multiplied by 4 and the organic half reaction must be multiplied by 3.

$$4 MnO_4^- + 16 H^+ + 12 e^- \rightarrow 4 MnO_2 + 8 H_2O$$

$$3 R\text{-}CH_2OH + 3 H_2O \rightarrow 3 R\text{-}CO_2H + 12 H^+ + 12 e^-$$

Simplify:

$$3 R\text{-}CH_2OH + 4 MnO_4^- + 4 H^+ \rightarrow 3 R\text{-}CO_2H + 4 MnO_2 + 5 H_2O$$

Change to basic solution by adding 4 OH^- to each side:

$$3 R\text{-}CH_2OH + 4 MnO_4^- + 4 H_2O \rightarrow 3 R\text{-}CO_2H + 4 MnO_2 + 5 H_2O + 4 OH^-$$

Simplify again:

$$3 R\text{-}CH_2OH + 4 MnO_4^- \rightarrow 3 R\text{-}CO_2H + 4 MnO_2 + H_2O + 4 OH^-$$

Since the carboxylic acid reacts with hydroxide ion, the last equation must be rewritten:

$$3 R\text{-}CH_2OH + 4 MnO_4^- \rightarrow 3 R\text{-}CO_2^- + 4 MnO_2 + 4 H_2O + OH^-$$

Finally, since $KMnO_4$ is used, add potassium(K^+) ions:

$$3 R\text{-}CH_2OH + 4 KMnO_4 \rightarrow 3 R\text{-}CO_2K + 4 MnO_2 + 4 H_2O + KOH$$

Experiment 4: CYCLOHEXANONE FROM CYCLOHEXANOL

OH

$Na_2Cr_2O_7$ / H_2SO_4 →

O

Cyclohexanol	Cyclohexanone
MW: 100.16	MW: 98.14
density: 0.96 g/mL	density: 0.95 g/mL
mp: 25 °C bp: 161 °C	mp: -31.2 °C bp: 156 °C
	solubility: 1.5 g/100 mL in H_2O at 10 °C

A stock solution of sodium dichromate ($Na_2Cr_2O_7.2H_2O$, 0.54 M) in sulfuric acid (2.52 M) will be available for you to use.[1] Transfer 6 g of cyclohexanol to a 125 mL Erlenmeyer flask containing 20 mL of distilled water.[2] Add 50 mL of the stock sodium dichromate solution to the flask containing the cyclohexanol.[3] Completely mix (using a glass stirring rod **not** a thermometer) the resulting solution.[4] Using a thermometer, measure the temperature of the solution. The reaction should be exothermic. Swirl occasionally, watch the temperature, and cool if necessary to keep the temperature between 55 and 60°C for about 15 min. Over the course of the reaction the color of the mixture should change from orange-brown to dark blue-green. Allow some time for additional reaction (10-15 min).[5] (During this time you can set up the distillation apparatus, take melting points, or wash glassware.)

Transfer the reaction mixture to a 250-mL round-bottomed flask using a funnel, rinse the Erlenmeyer flask with 75 mL of water, and add the water to the round-bottomed flask. Plug the side-arm of the flask with a cork, add a few boiling chips, and distill in a simple distillation apparatus (see the Figure labeled "Simple Distillation" in "Apparatus and Techniques"). (This is a steam distillation in which the steam is generated internally rather than from an outside source.) Initially, the distillate will be two phases (sometimes only one cloudy phase is seen), but gradually only water will be in the distillate. Note the temperature of the distillation. Collect about 50 mL of distillate in a graduated cylinder.

Because cyclohexanone is fairly soluble in water, simply extracting cyclohexanone from the distillate into an organic solvent may not be very efficient. By dissolving an inorganic salt such as sodium chloride in the aqueous layer, the solubility of cyclohexanone in the aqueous layer can be substantially reduced. This technique of specifically adding a salt to the aqueous layer is known as "salting out."

Accordingly, transfer the distillate to a 125 mL Erlenmeyer flask and then add 0.2 g of solid sodium chloride per milliliter of water present in the distillate and swirl to completely dissolve the salt. Then pour the mixture into a separatory funnel. Shake and vent as usual and separate the layers into two flasks. Label them organic (upper) and aqueous (lower). Make sure you know which is the organic layer and which is the aqueous layer.

Return the aqueous layer to the separatory funnel and add 20 mL of dichloromethane (methylene chloride). Shake and vent as usual, then drain the dichloromethane (lower) layer into the organic flask. If you did it right, the liquid in the organic flask will be homogeneous (no separate layers). Make sure you know which is the organic layer. Drain the aqueous (top) layer back into the aqueous flask.

Return the organic layer to the separatory funnel, and add an equal volume of saturated aqueous sodium chloride solution. Shake, vent, and drain the lower (dichloromethane) layer into a clean, dry flask. Save the upper aqueous layer for possible use later on. Now make sure you are working with the organic (dichloromethane) layer. Add 3-4 g anhydrous sodium sulfate to the flask containing the organic layer and swirl occasionally for 5 minutes in order to completely remove water from the solution. This solution, containing the organic layer and anhydrous sodium sulfate, can be saved until the next laboratory period by tightly stoppering the flask.

Decant the liquid into a dry 50 ml round-bottom flask, add 1 or 2 boiling chips, and perform a simple distillation (see the Figure labeled "Simple Distillation" in "Apparatus and Techniques"). Make sure the entire distillation apparatus is dry. Dichloromethane (bp 40°C) and cyclohexanone (bp 155-156°C) should be very easy to separate, but it is difficult to separate unreacted cyclohexanol (bp 161 °C) from cyclohexanone. You may wish to collect three fractions. The first fraction (bp 40-60°C) will be mostly dichloromethane. The next fraction (bp 60-140°C) should be very small. Collect the final fraction (usually 150-155°C) in a clean pre-weighed 18x150 mm test tube. Save a few drops for gas chromatography (GC) and infrared (IR) spectroscopy, and turn in the remainder. Note: if the boiling temperature of the distillate is around 100 °C, you are distilling the aqueous layer.

Before You Leave: Discard the green or blue green liquid residue from the steam distillation into the container marked "chromium waste". The aqueous layers from the extractions, and the used drying agent (Na_2SO_4) can be dissolved in water and poured down the sink with the water running. The residue from the simple distillation and the dichloromethane fraction should be discarded in the "organic waste" container.

Analysis of Product

Analyze your product by GC to determine its purity. If you collected more than one distillation fraction you may wish to determine the GC purity of both fractions

As usual, GC's should be labeled with your name and the GC conditions (temperature, time, program rate, etc.); all major peaks should be identified (name or structural formula). This information along with calculations of the % composition of your samples must be recorded in your notebook.

Take the IR spectrum of your product, and compare it to that of cyclohexanol. The IR spectrum should be labeled with your name, the name (or structural formula) of the compound, and the phase (liquid film, KBr pellet, or nujol mull). For this experiment, also label the major peaks (O-H, C-H, and C=O stretching vibrations). Record this analysis in your notebook.

Do not use your GC's and IR's as scratch paper. They should be turned in with your report for this experiment or pasted in your notebook, according to your instructor's directions.

The IR spectra of cyclohexanol and cyclohexanone are shown below.

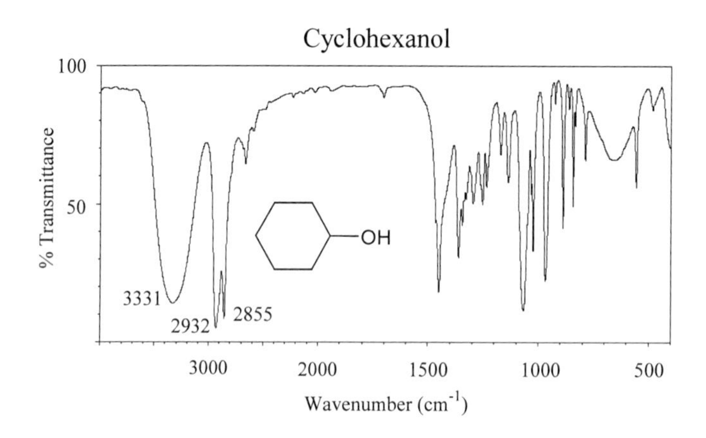

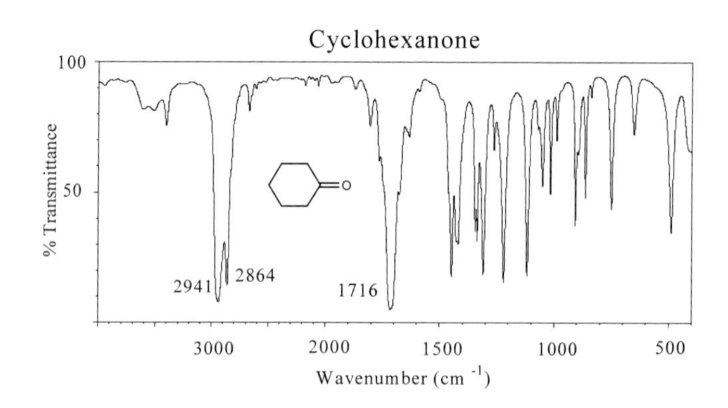

NAME Jarius Garner I.D. @02877197

YIELD REPORT EXPERIMENT 4: SYNTHESIS OF CYCLOHEXANONE FROM CYCLOHEXANOL

WEIGHT OF CYCLOHEXANOL	_________	g
MOLES OF CYCLOHEXANOL	_________	mol
VOLUME SODIUM DICHROMATE/SULFURIC ACID SOLUTION	_________	mL
MOLES OF SODIUM DICHROMATE	_________	mol
MOLES OF SULFURIC ACID	_________	mol
LIMITING REAGENT	_________	
THEORETICAL YIELD OF CYCLOHEXANONE	_________	mol
THEORETICAL YIELD OF CYCLOHEXANONE	_________	g
ACTUAL YIELD OF CYCLOHEXANONE	_________	g
PERCENT CYCLOHEXANONE IN DISTILLATE	_________	%
PERCENT YIELD OF CYCLOHEXANONE	_________	%
BOILING POINT (RANGE) OF PRODUCT	_________	°C

THIS PAGE SHOULD BE TURNED IN WITH YOUR SAMPLE. THE REQUIRED INFORMATION SHOULD ALSO BE RECORDED IN YOUR NOTEBOOK. THE LABEL ON EACH SAMPLE MUST INCLUDE YOUR NAME AND THE NAME OF THE COMPOUND.

Use of the Perkin Elmer Spectrum 100 FT-IR Spectrometer

Left click on NEW to clear an existing spectrum and get a blank spectrum window.

From the Instrument pull down menu choose SCAN. Then click on "Start". Click on "OK" if "warning duplicate file name" appears.

While on the "Monitoring Screen" place your sample on the window as indicated below.

Liquid Sample: Put one drop of the liquid over the window in the metal plate using a Pasteur pipette. Do not touch the window with the tip of the Pasteur pipette. Do NOT use the pressure arm. If the liquid is volatile (like cyclohexene), cover the liquid with a clean salt (NaCl) plate.

Solid Sample: Place a small amount of the solid over the window in the metal plate. Swing the pressure arm into place and start screwing it down (clockwise). When the pressure arm engages the sample, the Force Gauge on the monitor screen will start showing green. Continue screwing the pressure arm down until the Force Gauge shows a value of about 100. If the color changes to RED, STOP and then unscrew the pressure arm until the Force Gauge shows a value of about 100.

Click on SCAN

After the scan is complete, click on "Overwrite" two times in succession.

Click on PEAKS to have each peak labeled.

Click on Print to print out the IR spectrum.

Cleanup:

Liquid Sample: Blot up the liquid using a clean kimwipe. Then wipe the metal plate with a kimwipe that has been wetted with acetone or dichloromethane. Repeat the wiping 2 more times using a fresh kimwipe wetted with acetone or dichloromethane each time. If you used a salt plate to prevent evaporation of the liquid, clean the salt plate by washing it with acetone or dichloromethane.

Solid Sample: Unscrew the pressure arm (counterclockwise) and then rotate it away from the sample. Wipe up the sample using a clean kimwipe that has been wetted with acetone or dichloromethane. Then wipe the underside of the pressure arm with a kimwipe that has been wetted with acetone or dichloromethane. Repeat the wiping of the metal plate and the underside of the pressure arm 2 more times using a fresh kimwipe wetted with acetone or dichloromethane each time.

Do not attempt to turn off the FT-IR instrument or its computer.

ESTERIFICATION

A common method of preparing esters is the acid catalyzed reaction between a carboxylic acid and an alcohol. This process is called "Fischer" esterification. The acid catalyst acts to protonate the carboxylic acid, making the carboxylic carbon atom more susceptible to nucleophilic attack by an unshared pair of electrons on the oxygen atom of the alcohol. The mechanism of Fischer esterification (illustrated for the reaction of benzoic acid and methanol) is as follows:

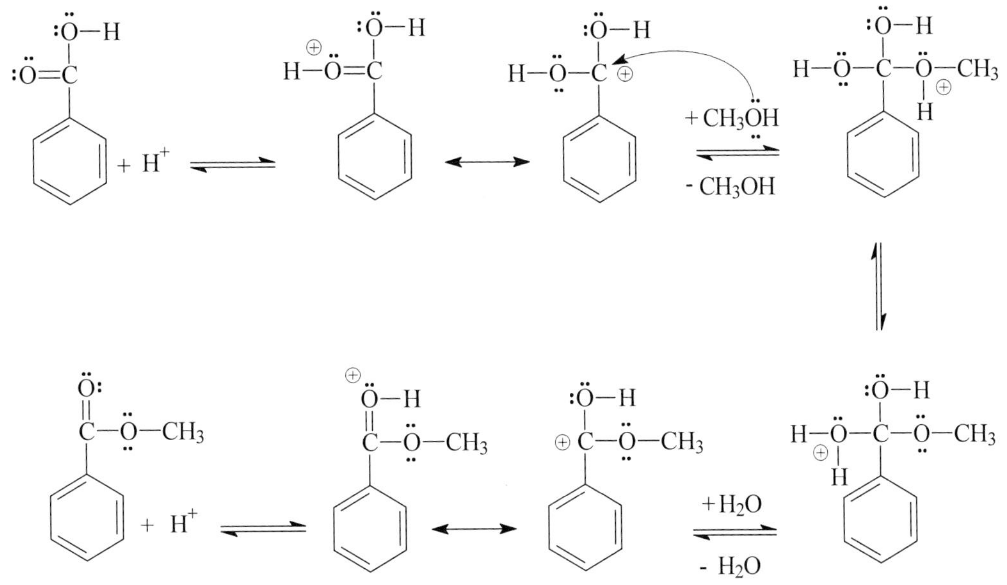

+ H^+ ⇌ ⟷ + CH_3OH ⇌ - CH_3OH

+ H^+ ⇌ ⟷ + H_2O ⇌ - H_2O

Each step of the mechanism is reversible. The reverse process is the mechanism for an acid catalyzed hydrolysis of an ester.

In Fischer esterification, the alcohol and the carboxylic acid are heated together in the presence of a strong acid catalyst such as sulfuric acid until equilibrium is achieved. The equilibrium constant for the reaction between benzoic acid and methanol to give methyl benzoate plus water is around 4.

$$K_{eq} = \frac{[C_6H_5CO_2CH_3]\ [H_2O]}{[C_6H_5CO_2H]\ [CH_3OH]} = 4$$

Thus if 1 mole of benzoic acid and 1 mole of methanol are reacted, the equilibrium mixture will contain 0.67 moles each of methyl benzoate and water and 0.33 moles each of benzoic acid and methanol. An efficient separation of this mixture would give methyl benzoate in an overall yield of 67%. According to LeChâtelier's principle the reaction can be driven towards the product side by adding an excess of one reactant or continuously removing one product.. Thus if 10 moles of methanol are reacted with 1 mole of benzoic acid, the equilibrium mixture will contain 0.97 moles of methyl benzoate, and the overall yield (based on benzoic acid as the limiting reagent, and assuming an efficient separation of the product) will be 97%. In many cases, such as the experiment you will carry out, the alcohol is used as the solvent, and the reaction is carried out at the boiling point of the solvent. Carrying out a reaction or other process at the boiling point of the solvent is known as **refluxing** (see “Refluxing” in “Apparatus and Techniques). In this experiment the molar ratio of methanol to benzoic acid is approximately 7.5:1, which means that at equilibrium the maximum yield of methyl benzoate is 96%.

Experiment 5: SYNTHESIS OF METHYL BENZOATE BY FISCHER ESTERIFICATION

$$C_6H_5COOH + CH_3OH \xrightleftharpoons{H_2SO_4} C_6H_5COOCH_3 + H_2O$$

Benzoic Acid	Methanol	Methyl benzoate
MW: 122.12	MW: 32.04	MW: 136.15
density: 1.316 g/mL	density: 0.791 g/mL	density: 1.087 g/mL
mp: 122.4 °C bp: 250°C	mp: -98 °C bp: 64 °C	mp: -13 °C bp: 198-199 °C

Weigh out 5.0 g of benzoic acid and transfer it to a 100-mL round-bottomed flask. Add 12.5 mL of methanol and cool the mixture in ice. Slowly and carefully add 1.5 mL of concentrated sulfuric acid by pouring it down the side of the flask (as opposed to directly into the solution). Then gently swirl to mix the reagents. Add one or two boiling chips and then attach a reflux condenser, and reflux the mixture gently for 1 h. with water slowly flowing through the condenser. See the Figure labeled "Refluxing" in "Apparatus and Techniques." Record both the time you started and stopped refluxing the mixture in your notebook. Cool the solution, then decant it into a separatory funnel containing 25 mL of water, and rinse the reaction flask with 20 mL of dichloromethane. Add the dichloromethane used to rinse the reaction flask to the separatory funnel. Shake the separatory funnel thoroughly, and then separate the layers. Discard the aqueous layer (containing the sulfuric acid and the majority of the methanol). Wash the organic layer (dichloromethane layer) in the separatory funnel with 15 mL of water. Separate the layers and discard the aqueous layer. Wash the organic layer (dichloromethane layer) with 15 mL of 5% sodium bicarbonate in order to remove any unreacted benzoic acid (production of CO_2 gas may occur). Again shake the separatory funnel with frequent release of pressure by inverting the separatory funnel and opening the stopcock, until no further reaction is apparent; then separate the layers. Make sure that the aqueous layer is basic (turns red litmus paper blue) by using a clean stirring rod to remove a drop of the aqueous layer and testing it with red litmus paper. If the aqueous layer is not basic, extract the organic layer (dichloromethane layer) with another 15 mL portion of 5% sodium bicarbonate. (Unreacted benzoic acid can be recovered by neutralizing the bicarbonate washes with 6M HCl.) Wash the organic (dichloromethane) layer in the separatory funnel with approximately 15 mL of saturated sodium chloride solution, and then dry the organic layer over anhydrous sodium sulfate in a 125 mL Erlenmeyer flask. Make sure you add

9:37 AM – 10:37 AM

enough anhydrous sodium sulfate so that the drying agent no longer clumps together on the bottom of the flask. After 10 min, decant the dry dichloromethane solution into a clean, dry 100 mL round-bottomed flask. Rinse the drying agent with an additional 5 mL of dichloromethane, and decant this rinse into the round-bottomed flask. Add one or two boiling chips to the round-bottomed flask.

Remove the dichloromethane by simple distillation (see the Figure labeled "Simple Distillation" in "Apparatus and Techniques"). Make sure that the apparatus you use is clean and dry. Collect the dichloromethane distillate in an Erlenmeyer flask. Once all the dichloromethane has distilled off, but before the methyl benzoate begins to distill, first turn off the water to the condenser, and then drain the condenser by removing the water inlet tubing from the water faucet. Because the boiling point of methyl benzoate is very high (199 °C), a water-cooled condenser could crack. Use a tared, clean, dry 18x150 mm test tube to collect the material that boils above 190 °C. Be sure to record in your notebook the temperature range for the boiling point of the material you actually collect. A small amount of solution must remain in the round-bottomed flask (do not distill to dryness).

Before You Leave: Combine all the aqueous layers and then neutralize this solution with sodium carbonate. The neutralized solution can be poured down the drain followed by lots of water. If the sodium sulfate is free of dichloromethane and methyl benzoate, it can be placed in the nonhazardous solid waste container or dissolved in water and then poured down the drain followed by lots of water; otherwise it must go into the hazardous solid waste container. The dichloromethane from the distillation is added to the waste organic container, along with the residues from the final distillation.

Analyze the methyl benzoate by GC, IR, and TLC. Paste/staple the GC and IR outputs in your notebook. Record the GC conditions and your analysis of both the GC and the IR in your notebook.

For the TLC analysis, prepare your developing chamber as before but using a mixture of dichloromethane : ligroin (50:50) as your mobile phase. Again, your stationary phase is silica gel ($SiO_2 \cdot xH_2O$) coated on a thin sheet of aluminum. For this experiment, you will spot a solution of your isolated methyl benzoate along with a solution of benzoic acid in dichloromethane.

Prepare a solution of each sample (~1 mg) in a few drops of dichloromethane or fill the narrow end of a Pasteur pipette to a depth of about 1.0 cm with your solid or liquid sample. Dissolve this amount of each sample in 8-10 drops of dichloromethane in separate small test tubes. Make a pencil mark about 6 mm from the bottom of the plate (do NOT use a pen). Using a fine capillary tube or pipette tip, spot each sample solution as close to this pencil line as possible. (Be careful not to make a large spot, and make sure that your spot is above the solvent line once you insert the plate in the developing chamber). Make a second pencil mark about 3-4 mm from the top of the plate. The stationary phase for this experiment is silica gel ($SiO_2 \bullet xH_2O$) coated on a thin sheet of aluminum.

Develop your plate in the developing chamber, so that the solvent travels up the plate by capillary action carrying components of the mixture along the silica gel stationary phase. Remove the plate when the solvent just reaches the second pencil line that is about 3-4 mm from the top of the plate. This line is the position of your solvent front. Wait about 30-40 seconds to allow the solvent to completely evaporate from the plate and then observe the plate under a UV lamp (Do NOT look directly at the UV light). Make note of the position of all spots seen under the UV lamp and use a pencil to outline each spot. Calculate the R_f value for your methyl benzoate and comment on (a) the purity of your isolated methyl benzoate and (b) the relative affinity of benzoic acid and methyl benzoate for the stationary phase. Make a sketch of each of your developed TLC plate in your laboratory notebook that shows the relative positions of each spot seen under the UV lamp. Also attach the TLC plate with the spots outlined in pencil in your notebook.

The IR spectra of benzoic acid and methyl benzoate are shown below.

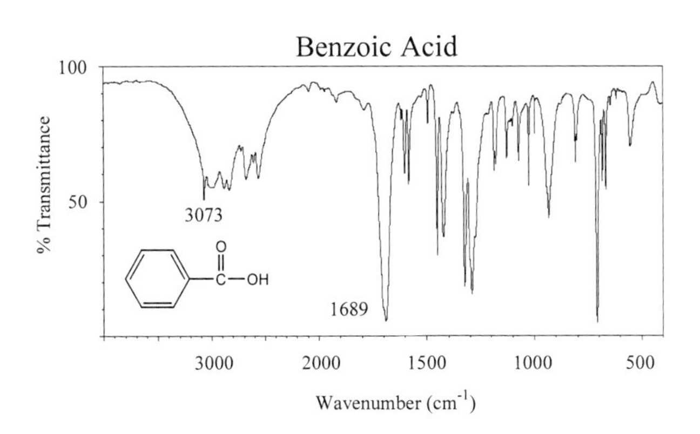

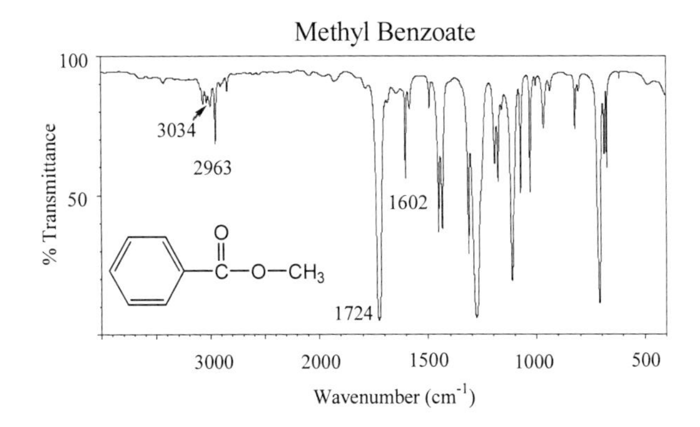

NAME Jarius Garner I.D. @02877197

YIELD REPORT EXPERIMENT 5: SYNTHESIS OF METHYL BENZOATE

WEIGHT OF BENZOIC ACID	5.030	g
MOLES OF BENZOIC ACID	0.04119	mol
VOLUME OF METHANOL	12.5	mL
MOLES OF METHANOL	0.0039	mol
LIMITING REAGENT	Benzoic acid	
THEORETICAL YIELD OF METHYL BENZOATE	~~0.041~~	mol
THEORETICAL YIELD OF METHYL BENZOATE	~~5.61~~	g
ACTUAL YIELD OF METHYL BENZOATE	-	g
PERCENT YIELD OF METHYL BENZOATE	-	%
BOILING POINT (RANGE) OF PRODUCT	160 - 170	°C

THIS PAGE SHOULD BE TURNED IN WITH YOUR SAMPLE. THE REQUIRED INFORMATION SHOULD ALSO BE RECORDED IN YOUR NOTEBOOK. THE LABEL ON EACH SAMPLE MUST INCLUDE YOUR NAME AND THE NAME OF THE COMPOUND.

Theoretical yield

NITRATION OF METHYL BENZOATE

Methyl benzoate as well as other aromatic compounds can be nitrated using a mixture of nitric and sulfuric acids. The actual nitrating reagent is the positively charged nitronium ion (NO_2^+) which is formed in the reaction mixture by the dehydration of nitric acid facilitated by sulfuric acid.

$$HNO_3 + 2\,H_2SO_4 \rightleftarrows NO_2^+ + 2\,HSO_4^- + H_3O^+$$

In this experiment sulfuric acid also is the solvent. As illustrated below, sulfuric acid is a strong enough acid that it can protonate and thus dissolve methyl benzoate:

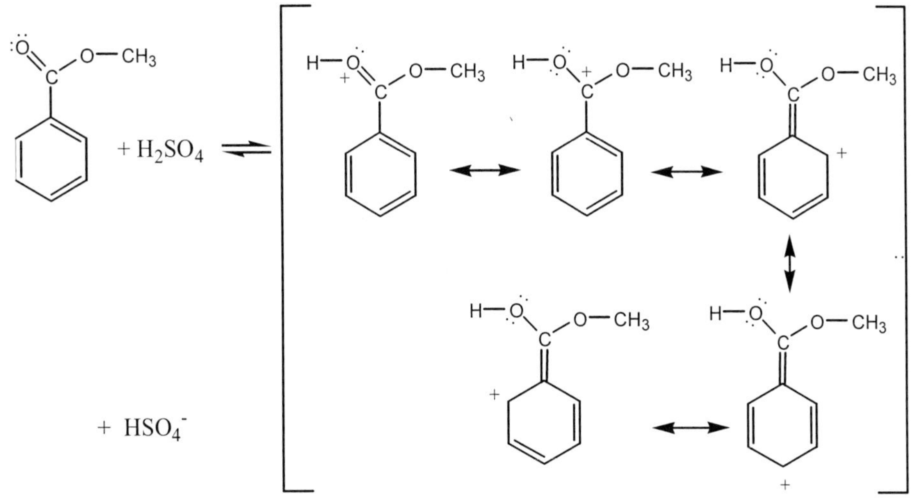

$+ H_2SO_4 \rightleftharpoons$

$+ HSO_4^-$

Nitration at the meta position is favored because the partial positive charges residing at the ortho and para positions repel the positively charged nitronium ion.

Although the carbomethoxy ($-CO_2CH_3$) group is *meta*-directing, the actual product contains *ortho* and *para* isomers as well. The reported isomer distribution (for the ethyl ester, which should be similar to the methyl ester) for nitration is shown in Table 1.

Table 1. Nitration of methyl benzoate and benzoic acid

Isomer	**% of Product**	**mp (methylnitro benzoate) °C**	**mp (nitro benzoic acid) °C**
ortho	24-28	-8	147-148
meta	66-73	78	140-142
para	1-6	96	242

The *para* isomer is a very minor product, but the *ortho* isomer is about a quarter of the nitration product. In the nitration of methyl benzoate, the *ortho* isomer (methyl 2-nitrobenzoate) is not isolated and should not appreciably contaminate the *meta* product (methyl 3-nitrobenzoate) because the *ortho* isomer is a liquid at room temperature. The *meta* isomer (methyl 3-nitrobenzoate) is a solid which precipitates from the reaction mixture (see the Table 1 for the melting points of the isomeric esters). Nitration of benzoic *acid* gives approximately the same isomer distribution, but isolation of the pure *meta* isomer (3-nitrobenzoai acid) is complicated because the *ortho* (2-nitrobenzoic acid) and *meta* (3-nitrobenzoic acid) isomers are both solids with similar melting points (see Table 1). The best way to obtain pure 3-nitrobenzoic acid would be nitration of the methyl ester, as in this experiment, followed by hydrolysis of the ester.

CO_2H CH_3OH H_2SO_4 CO_2CH_3

HNO_3 + H_2SO_4

CO_2H H_2O H_2SO_4 CO_2CH_3

NO_2 NO_2

Experiment 6: NITRATION OF METHYL BENZOATE

Methyl benzoate
MW: 136.15
density: 1.087 g/mL
mp: -13 °C bp: 198-199 °C

$\xrightarrow[H_2SO_4]{HNO_3}$

Methyl 3-nitrobenzoate
MW: 181.15
mp: 78 °C bp: 279 °C

Transfer 6 mL of concentrated sulfuric acid to a 125-mL Erlenmeyer flask and then cool the solution to about 0 °C in an ice bath. Add 3.0 g of methyl benzoate to the ice cold sulfuric acid. Cool the mixture in the ice bath so that its temperature is 0-10 °C. Prepare a cooled (0 to 10 °C) mixture containing 2 mL of concentrated sulfuric acid and 2 mL of concentrated nitric acid in another Erlenmeyer flask. Using a Pasteur pipette add dropwise the cooled nitric/sulfuric acid mixture to the cooled sulfuric acid solution of methyl benzoate. Gently swirl the methyl benzoate solution after addition of each few drops of the nitric/sulfuric acid mixture. Do not let the temperature of the reaction mixture get above 15 °C.

After all the nitric/sulfuric acid has been added, allow the reaction mixture to warm up to room temperature and then wait 15 min. Pour the reaction mixture onto 25 g of cracked ice in a 100 mL beaker. Isolate the solid product by vacuum filtration using a small Buchner funnel (see "Vacuum Filtration" in "Apparatus and Techniques") and wash the solid well, first with ice cold water (do not get any ice in the Büchner funnel) and then with two 7-mL portions of ice-cold methanol. Finally air dry the solid (crude product). Save a small sample of the crude product for a melting point determination and analysis by TLC and GC. The remaining crude product is weighed and recrystallized from an equal weight of methanol in a 6" test tube. Use a hot water bath (1000 mL beaker filled with about 700 mL water on a hot plate heated to ~70 °C) to heat the test tube until all the solid dissolves. Isolate the recrystallized product by vacuum filtration. Wash the crystals with a few mL of ice cold methanol and then air dry them completely.

Determine the melting point of your crude and recrystallized products and analyze them both by GC and TLC (see below). Obtain an IR of the recrystallized product. After you have completed the analysis of your sample of methyl 3-nitrobenzoate, turn your recrystallized sample in to your instructor in a labeled test tube. The crude methyl 3-nitrobenzoate will have a melting point of about 74-76 °C. The melting point of the recrystallized methyl 3-nitrobenzoate should be 78 °C.

Before You Leave: Dilute the filtrate from the reaction with water, neutralize it with sodium carbonate, and flush the neutralized solution down the drain followed by lots of water. The methanol from the crystallization should be placed in the waste organic container.

For the TLC analysis, prepare your developing chamber as before but using a mixture of dichloromethane : ligroin (50:50) as your mobile phase. For this experiment, you will spot solutions of your crude product, your recrystallized product, and methyl benzoate, each disolved in dichloromethane.

Prepare a solution of each sample (~1 mg) in a few drops of dichloromethane or fill the narrow end of a Pasteur pipette to a depth of about 1.0 cm with your sample. Dissolve this amount of your solid or liquid sample in 8-10 drops of dichloromethane in separate small test tubes. Make a pencil mark about 6 mm from the bottom of the plate (do NOT use a pen). Using a fine capillary tube or pipette tip, spot each sample solution as close to this pencil line as possible. (Be careful not to make a large spot, and make sure that your spot is above the solvent line once you insert the plate in the developing chamber). Make a second pencil mark about 3-4 mm from the top of the plate. The stationary phase for this experiment is silica gel ($SiO_2 \bullet xH_2O$) coated on a thin sheet of aluminum.

Develop your plate in the developing chamber, so that the solvent travels up the plate by capillary action carrying components of the mixture along the silica gel stationary phase. Remove the plate when the solvent just reaches the second pencil line that is about 3-4 mm from the top of the plate. This line is the position of your solvent front. Wait about 30-40 seconds to allow the solvent to completely evaporate from the plate and then observe the plate under a UV lamp (Do NOT look directly at the UV light). Make note of the position of all spots seen under the UV lamp and use a pencil to outline each spot. Calculate the R_f values of all the spots observed, making note of the number of spots observed in your crude and your recrystallized products. Make a sketch of your developed TLC plate in your laboratory notebook that shows the relative positions of each spot seen under the UV lamp. Also attach the TLC plate with the spots outlined in pencil in your notebook.

The IR spectra of methyl benzoate and methyl 3-nitrobenzoate are shown below.

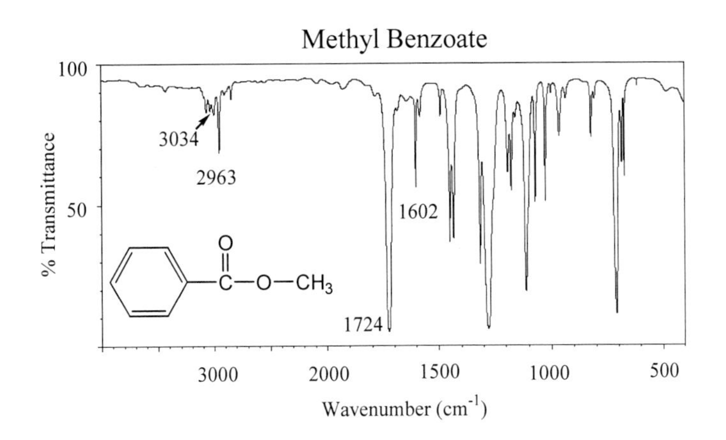

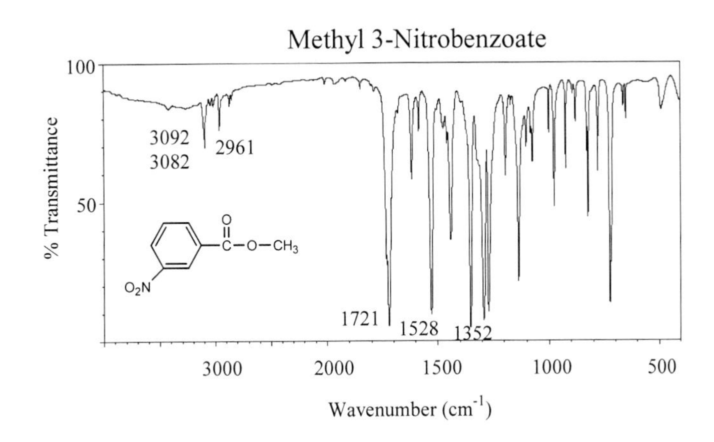

NAME_______________________________________ I.D.______________

YIELD REPORT EXPERIMENT 6: NITRATION OF METHYL BENZOATE

WEIGHT OF METHYL BENZOATE __________ g

MOLES OF METHYL BENZOATE __________ mol

VOLUME OF NITRIC ACID __________ mL

MOLES OF NITRIC ACID __________ mol

VOLUME OF SULFURIC ACID __________ mL

MOLES OF SULFURIC ACID __________ mol

THEORETICAL YIELD OF METHYL 3-NITROBENZOATE __________ mol

THEORETICAL YIELD OF METHYL 3-NITROBENZOATE __________ g

ACTUAL YIELD OF CRUDE METHYL 3-NITROBENZOATE __________ g

WEIGHT OF CRUDE METHYL 3-NITROBENZOATE USED IN
RECRYSTALLIZATION __________ g

ACTUAL YIELD OF RECRYSTALLIZED
METHYL 3-NITROBENZOATE __________ g

PERCENT RECOVERY FOR RECRYSTALLIZATION __________ %

PERCENT YIELD OF METHYL 3-NITROBENZOATE __________ %

MELTING POINT (RANGE) OF CRUDE PRODUCT __________ °C

MELTING POINT (RANGE) OF RECRYSTALLIZED PRODUCT __________ °C

THIS PAGE SHOULD BE TURNED IN WITH YOUR SAMPLE. THE REQUIRED INFORMATION SHOULD ALSO BE RECORDED IN YOUR NOTEBOOK. THE LABEL ON EACH SAMPLE MUST INCLUDE YOUR NAME AND THE NAME OF THE COMPOUND.

THE GRIGNARD SYNTHESIS

Grignard reagents belong to a large class of organic compounds that are called organometallic reagents. The general formula for these compounds is R-M, where M is a metal (e.g., Li, Na, Mg, Cd, Al, Cu, etc). Accordingly, these reagents contain a carbon-metal (C-M) bond, and since carbon is more electronegative than a metal, the polarity of the C-M bond in these compounds is as shown below, with carbon being

$$\overset{\delta^-}{C}—\overset{\delta^+}{M}$$

the electron rich (nucleophilic) end of the bond. Grignard reagents, whose general formula is RMgX (X = Cl, Br, or I), contain a carbon-magnesium bond. Because of the large difference in the electronegativities of carbon and magnesium, the C-Mg bond has a large amount of ionic character, hence Grignard reagents are very reactive compounds and play an immensely important role in organic syntheses.

Grignard reagents are prepared from organic halides by the reaction shown in eqn 1, where R = alkyl, vinyl or aryl, X = Cl, Br, or I, and Ether is an ether, usually diethyl

$$RX + Mg \xrightarrow{\text{Ether}} RMgX \qquad (1)$$

ether or tetrahydrofuran. While this reaction maybe viewed simplistically as an insertion reaction (i.e., insertion of the Mg into the C-X bond of R-X), it can also be viewed as an oxidation (of Mg)/reduction (of R-X) reaction. The reaction takes place on the surface of the metal; thus, in order to increase the rate of this reaction, Mg turnings (large surface area), rather than a chunk (small surface area) of Mg metal, are used. The ether used in this reaction plays a dual role of both the solvent and a reactant. As a reactant, the ether functions as a Lewis base and solvates the Grignard reagent through complexation (eqn 2) as soon as the Grignard reagent is produced on the surface of the metal. This complex then diffuses away from the surface of the Mg, freeing the surface up for further reaction.

$$RMgX + 2\ \underset{\text{Ether}}{R'—\ddot{O}—R'} \longrightarrow R'_2O \cdots RMgX \cdots OR'_2 \qquad (2)$$

Grignard reagents are strong bases, therefore, they must be prepared and used under anhydrous conditions. In the presence of even weak acids (e.g., water, alcohols), the reaction shown in eqn 3 occurs instantaneously. Thus, all of the glassware and chemicals that are used in this reaction must be completely dry (anhydrous).

$$RMgX + HOH \longrightarrow RH + Mg(OH)X \qquad (3)$$

While Grignard reagents are also good nucleophiles and in some cases can react with the starting organic halide to give a coupling product as shown in eqn 4, mechanistic studies have shown that radicals (R•) are formed

$$RMgX + RX \longrightarrow R\text{—}R + MgX_2 \qquad (4)$$

during the preparation of the Grignard reagent. Thus, this coupling product might also be formed by the combination of two radicals, as shown in eqn 5. In any case, the

$$R\bullet + R\bullet \longrightarrow R\text{—}R \qquad (5)$$

formation of this by-product (R-R) becomes important if the concentration of RX in the reaction mixture is allowed to get too high. Because of the relatively high ionic nature of the C-Mg bond, Grignard reagents are readily oxidized by air to peroxide anions as shown in eqn 6. Thus, these reagents cannot be stored easily and must be used soon after preparation.

$$RMgX + O_2 \longrightarrow ROO^- \; {}^+MgX \qquad (6)$$

Organic chemists take advantage of the reactivity of Grignard reagents with carbonyl compounds to synthesize alcohols (primary, secondary, and tertiary) and carboxylic acids. The type of alcohol formed is a function of the type of carbonyl compound used. Grignard reagents react: (i) with formaldehyde to give primary alcohols, eqn 7; (ii) with any other aldehyde to give secondary alcohols, eqn 8; (iii) with ketones to give tertiary alcohols, eqn 9; and (iv) with esters to give tertiary alcohols, eqn 10. (Note the differences in the structures of the two tertiary alcohols that are produced in reactions 9 and 10). Grignard reagents also react with carbon dioxide to give carboxylic acids, eqn 11.

$$RMgX + H-\overset{\overset{O}{\|}}{C}-H \xrightarrow{Ether} \xrightarrow[HX]{dil} R-CH_2-OH + MgX_2 \quad (7)$$

$$RMgX + R'-\overset{\overset{O}{\|}}{C}-H \xrightarrow{Ether} \xrightarrow[HX]{dil} R'-\underset{\underset{R}{|}}{\overset{\overset{OH}{|}}{C}}-H + MgX_2 \quad (8)$$

$$RMgX + R'-\overset{\overset{O}{\|}}{C}-R'' \xrightarrow{Ether} \xrightarrow[HX]{dil} R'-\underset{\underset{R}{|}}{\overset{\overset{OH}{|}}{C}}-R'' + MgX_2 \quad (9)$$

$$2\ RMgX + R'-\overset{\overset{O}{\|}}{C}-OR'' \xrightarrow{Ether} \xrightarrow[HX]{dil} R'-\underset{\underset{R}{|}}{\overset{\overset{OH}{|}}{C}}-R + 2\ MgX_2 \quad (10)$$

$$RMgX + O=C=O \xrightarrow{Ether} \xrightarrow[HX]{dil} R-\overset{\overset{O}{\|}}{C}-OH + MgX_2 \quad (11)$$

The C=O bond in a carbonyl compound is polarized as shown below, with carbon being the electron deficient (electrophilic) end of the bond. (Note that the polarity of this bond

$$\overset{\delta^+}{C}=\overset{\delta^-}{O}$$

is the reverse of that of the C-Mg bond). Thus, all of these reactions involve the nucleophilic addition of the Grignard reagent to the carbonyl carbon of the carbonyl compound. When an ester is used (eqn 10), R" is usually a methyl or an ethyl group. This is done for practical reasons, one of which is to make sure that the by-product alcohol, R"OH, will be water soluble, hence easily separated from the desired tertiary alcohol product.

In this experiment, you will prepare phenylmagnesium bromide from bromobenzene according to the reaction shown in eqn 1, where R = phenyl, X = Br and Ether is anhydrous diethyl ether. Because of the procedure that you will use in this preparation, some of the coupling product (biphenyl, in this case) will be formed (see eqns 4 and 5, where R = phenyl and X = Br). You will then allow the phenylmagnesium bromide to react with methyl benzoate which gives, after acidification with dilute

$$2\ C_6H_5\text{-}MgBr + C_6H_5\text{-}\overset{O}{\overset{\|}{C}}\text{-}OCH_3 \xrightarrow{\text{Ether}} \xrightarrow[H_2SO_4]{\text{dil}} (C_6H_5)_3C\text{-}OH \qquad (12)$$

$$+\ CH_3OH\ +\ 2\ Mg(HSO_4)Br$$

aqueous acid, triphenylmethanol (eqn 12).

The mechanism by which this reaction occurs involves the following steps. In the first step, nucleophilic addition of the Grignard reagent to the carbonyl group of the ester gives an unstable addition product (eqn 13), which in the second step undergoes loss of an alkoxide ion (methoxide ion, in this case) to give a ketone (benzophenone, in this case) (eqn 14). In the third step, a second mole of the Grignard reagent adds to the carbonyl group in this ketone to give a new stable addition product, an alkoxide ion (eqn 15). At this point the Grignard reaction is over and the solution in highly basic. Acidification of the solution with some acid (dilute H_2S0_4, in this case) gives the final products shown in eqn 16.

At this point, in addition to the organic products shown in eqn 16, your reaction mixture may also contain benzene (formed by reaction of any unreacted phenylmagnesium bromide with the aqueous H_2SO_4, eqn 3), biphenyl (see eqns 4 and 5), as well as unreacted bromobenzene and unreacted methylbenzoate. By taking advantage of differences in their physical properties, these by-products will be removed during the purification of your triphenylmethanol. In particular, benzene and biphenyl are both hydrocarbons and are soluble in hydrocarbon solvents such as ligroin (a mixture of mostly hexanes and heptanes with a boiling range of 66-77 °C), where as triphenylmethanol is not soluble in ligroin.

δ^{-} δ^{+} MgBr + O, C—OCH_3 → O^{-} $^{+}MgBr$, C—OCH_3 (13)

O^{-} $^{+}MgBr$, C—OCH_3 → O, C + CH_3O^{-} $^{+}MgBr$ (14)

δ^{-} δ^{+} MgBr + O, C → $^{+}MgBr$, O^{-}, C (15)

$^{+}MgBr$, O^{-}, C + CH_3O^{-} $^{+}MgBr$ $\xrightarrow{H_2SO_4}$ OH, C (16)

+ CH_3OH + $Mg(HSO_4)Br$

Experiment 7: GRIGNARD SYNTHESIS OF TRIPHENYLMETHANOL

The following apparatus should be cleaned and allowed to dry in the lab period prior to the Grignard synthesis.

250-mL round-bottomed flask with side arm
125-mL separatory funnel
reflux condenser
Claisen adapter
vacuum adapter
small graduated cylinder
standard-taper glass stopper

Bromobenzene + Mg → (anhydrous ether) Phenylmagnesium bromide

Bromobenzene
MW 157.02
density: 1.495 g/mL
mp: -31 °C bp: 156 °C

Magnesium
At Wt 24.31

Phenylmagnesium bromide
not isolated, used *in situ*

2 Phenylmagnesium bromide + Methyl benzoate → Triphenylmethanol + CH_3OH

Phenylmagnesium bromide
not isolated, used *in situ*

Methyl benzoate
MW: 136.15
density: 1.087 g/mL
mp: -13 °C bp: 198-199 °C

Triphenylmethanol
MW: 260.34
mp: 163 °C bp: >360 °C

The reaction is carried out in a 250-mL round-bottomed flask with a side arm. Weigh 1.0 g of magnesium turnings, mash them in a mortar to expose the surface of the metal, and then immediately place the magnesium in the flask. Do **not** handle the magnesium with your fingers. Clamp the flask securely at the neck, high enough to allow space for an ice bath to be used if necessary. Cork the side arm, and attach the Claisen adapter to the flask, but do not grease the joint. Attach the 125-mL separatory funnel, which acts as an addition funnel, to the Claisen adapter directly above the flask, and a reflux condenser to the other joint of the Claisen adapter. Stopper the addition funnel.

To the top of the West Condenser attach a vacuum adapter packed with anhydrous calcium chloride and having a rubber bulb over its sidearm. See the figure below for a complete picture of the assembled apparatus. Do not start the water flowing in the condenser yet. Have a watery ice bath ready in case the reaction becomes too vigorous.

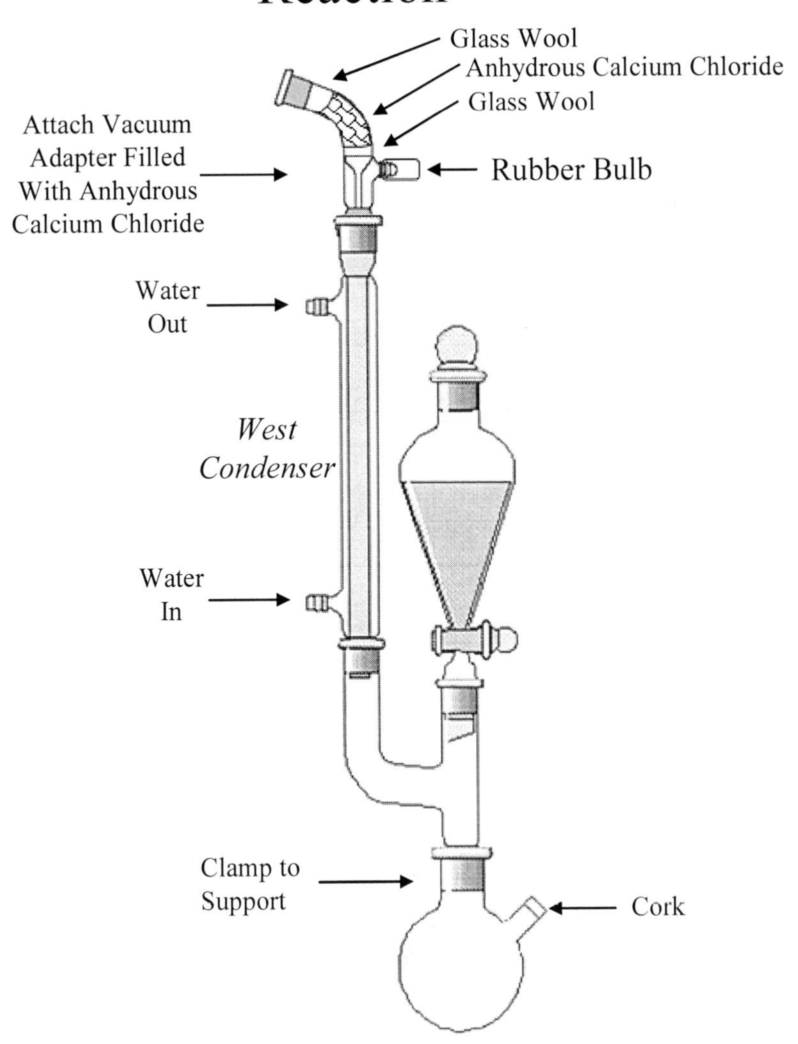

Preparation of phenylmagnesium bromide. Add 5.0 mL of bromobenzene and 10 mL of anhydrous diethyl ether to the separatory funnel and mix the liquids. Let the bromobenzene-diethyl ether solution into the flask by opening the stopcock. After the mixture is added, close the stopcock and place 15 mL of anhydrous diethyl ether in the separatory funnel.

When the reaction begins, the mixture will begin to get warm and cloudy and bubbles may form on the magnesium. If no change is observed after a few minutes, try the following to start the reaction, in this order..

1. Warm the flask by rubbing the bottom of the flask with the palm of your hand for five to ten minutes.
2. Carefully insert a clean dry glass stirring rod through the side arm and gently crush the metal under the liquid surface. Remove the stirring rod and cork the side arm, then continue rubbing the flask with the palm of your hand. Be careful not to puncture the flask with the stirring rod.
3. If no reaction begins after several more minutes, ask the instructor for help. A small quantity of 1,2-dibromoethane or a solution of a Grignard reagent made from 1,2-dibromoethane will be added to the reaction flask. The 1,2-dibromoethane will react with the magnesium in the flask and give it a clean surface. The Grignard reagent from 1,2-dibromoethane is unstable and immediately decomposes:

$$Br\text{-}CH_2CH_2\text{-}Br + Mg \rightarrow Br\text{-}CH_2CH_2\text{-}MgBr \rightarrow CH_2{=}CH_2 + MgBr_2$$

When the reaction begins, the mixture may turn dark brown and the ether will boil. At this point, start the water flowing slowly in the reflux condenser, and add the 15 mL of ether from the separatory funnel at such a rate that the mixture continues to reflux gently. If the boiling becomes too vigorous, cool briefly in an ice bath until a gentle reflux is obtained again. Do not over-cool.

The reaction is over when refluxing stops and only a few bits of magnesium remain in the reaction flask. Check the volume of the reaction mixture; if most of the ether has boiled away you may wish to add more anhydrous diethyl ether. Allow this mixture to stand a few minutes while you prepare for the next step. **Use this Grignard reagent today!**

Triphenylmethanol (triphenylcarbinol). Prepare a solution of 3.0 g of methyl benzoate (from stockroom) in 10 mL of anhydrous diethyl ether and add this solution to the separatory funnel. Slowly add this solution to the Grignard reagent, at a rate that maintains a gentle reflux. Slight cooling and swirling may be desirable. The reaction mixture will turn very cloudy and may be brown, orange, or pink. When all of the methyl benzoate solution has been added, and the reaction mixture has cooled to room temperature, disassemble the apparatus, cork the reaction flask, and allow the mixture to stand until the next laboratory period.

!!!STOP!!!

Pour the reaction mixture into about 25 ml of 10% sulfuric acid and 25 g of ice in a 250 mL beaker. Be prepared for an exothermic reaction. Foaming may occur. If the mixture has been standing for several days, the magnesium salt of triphenylmethanol will have probably crystallized out and the solid may be difficult to dissolve. If so, add dilute sulfuric acid to the reaction flask. It may be necessary to break up the lumps with a spatula or stirring rod and to also add more diethyl ether. (From here on the diethyl ether does not need to be anhydrous.) Rinse the reaction flask with dilute acid and diethyl ether, and add these washings to the separatory funnel. (Be sure the magnesium chips are dissolved before adding the liquid to the separatory funnel.) Separate the layers.

Wash the organic layer with about 25 mL of saturated NaCl solution. Dry the organic layer over anhydrous sodium sulfate (Na_2SO_4), and decant the dried organic layer into a 125 mL Erlenmeyer flask.

Add 25 mL of ligroin (66-77 °C) and concentrate the solution on a hot plate in the hood until most of the diethyl ether has been removed, or until crystals (triphenylmethanol) just start to appear. Remove the flask from the hot plate and allow the flask to cool to room temperature (crystallization may be slow, so you may want to let it stand for a few days.), then cool the flask in an ice bath. Using a Pasteur pipette carefully remove and then save a few mL of the supernatant liquid, the “mother liquor” for later GC analysis. Filter off the crystals (Büchner funnel/vacuum filtration). The crystals should be washed with a small volume (5 to 10 mL) of ice cold ligroin and then air dried. (A second crop of crystals can be obtained by further concentration of the filtrate and cooling again.) Weigh the crystals, determine their mp, analyze them by GC, TLC (see below), and IR, and turn them in. The GC outputs, IR spectrum, and TLC plates should be stapled or pasted into your notebook.

1.334 g

Analyze the remaining liquid (the "mother liquor") by GC (add 2 drops of the "mother liquor" to 10 drops of dichloromethane) and TLC (see below). If time permits and your instructor tells you, evaporate the "mother liquor" to dryness and weigh the residue. From the GC of the "mother liquor" estimate the ratio of methyl benzoate, biphenyl, triphenylmethanol, and unreacted bromobenzene present.

Before You Leave Any magnesium salts remaining in your glassware can be easily removed by washing with dilute acid. Remove the calcium chloride from the vacuum adapter. The calcium chloride should be discarded in the solid waste container. (If you leave the vacuum adapter full, the calcium chloride will initially cake up and become useless, and eventually absorb so much atmospheric moisture that it will liquefy and get your drawer messy.). Discard the ligroin filtrate in the Organic Waste container. After all the diethyl ether has evaporated, the solid sodium sulfate can be discarded in the solid waste container.

For this experiment, you will analyze the “mother liquor” and recrystallized product on the same TLC plate using two different solvent systems.

Plate 1: Develop the first plate using dichloromethane as the mobile phase
Plate 2: Develop a second plate using a mixture of dichloromethane/ligroin (50:50) as the mobile phase.

Fill the narrow end of a Pasteur pipette to a depth of about 1.0 cm with your recrystallized sample. Dissolve this amount of solid in 8-10 drops of dichloromethane in a small test tube. The "mother liquor" can be spotted directly. You do not need to prepare a sample solution of the "mother liquor".

Make a pencil mark about 6 mm from the bottom of each plate (do NOT use a pen). Using a fine capillary tube or pipette tip, spot each sample solution as close to this pencil line as possible. (Be careful not to make a large spot and make sure that your spot is above the solvent line once you insert the plate in the developing chamber). Make a second pencil mark about 3-4 mm from the top of each plate. Your stationary phase for this experiment is silica gel ($SiO_2 \bullet xH_2O$) coated on a thin sheet of aluminum.

Develop each plate in the developing chamber, so that the solvent travels up the plate by capillary action carrying components of the mixture along the silica gel stationary phase. Remove the plate when the solvent just reaches the second pencil line that is about 3-4 mm from the top of the plate. This line is the position of your solvent front.

Observe both plates under the UV lamp (do NOT look directly at the UV light), and mark an outline of each spot with a pencil. Then place each plate in a chamber of iodine for about 5 mins. Outline any additional spots that develop with a pencil. Calculate the R_f value of your recrystallized triphenylmethanol, and determine if some triphenylmethanol could be observed in your "mother liquor." Compare the two plates determining the number of components visible in the "mother liquor" with each mobile phase. Make a sketch of each of your developed TLC plates in your laboratory notebook that shows the relative positions of each spot seen under the UV lamp or detected by iodine. Indicate any spots that were only detected by iodine. Also Attach each TLC plate with the spots outlined in pencil in your notebook.

The IR spectra of bromobenzene, methyl benzoate, and triphenylmethanol are shown below.

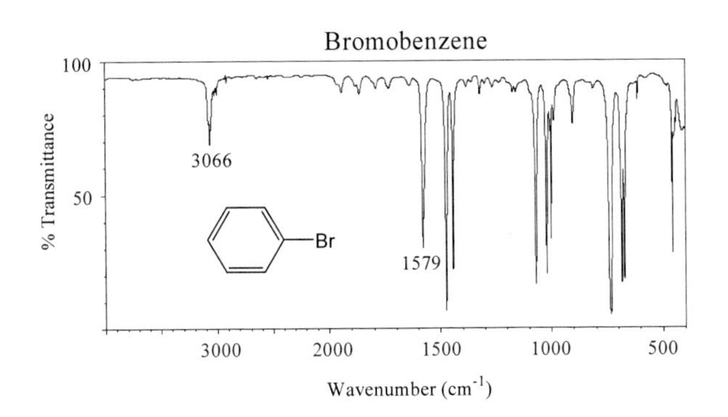

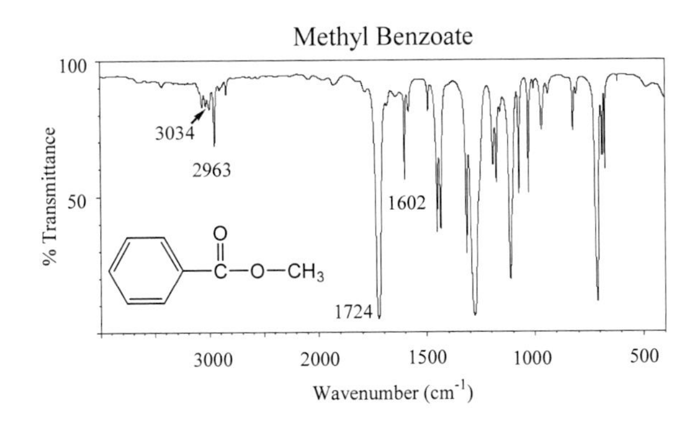

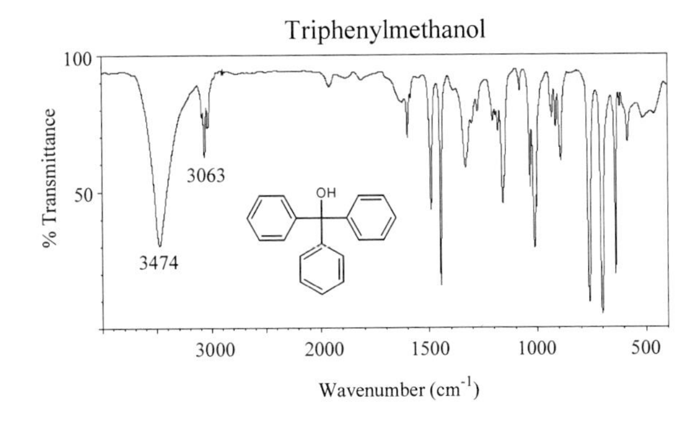

NAME Jarius Garner I.D.______________

YIELD REPORT EXPERIMENT 7: GRIGNARD SYNTHESIS OF TRIPHENYLMETHANOL

WEIGHT OF MAGNESIUM	__________ g
MOLES OF MAGNESIUM	__________ mol
WEIGHT OF BROMOBENZENE	__________ g
MOLES OF BROMOBENZENE	__________ mol
WEIGHT OF METHYL BENZOATE	__________ g
MOLES OF METHYL BENZOATE	__________ mol
LIMITING REAGENT	__________
THEORETICAL YIELD OF TRIPHENYLMETHANOL	__________ mol
THEORETICAL YIELD OF TRIPHENYLMETHANOL	__________ g
ACTUAL YIELD OF TRIPHENYLMETHANOL	__________ g
PERCENT YIELD OF TRIPHENYLMETHANOL	__________ %
MELTING POINT (RANGE) OF PRODUCT	__________ °C

THIS PAGE SHOULD BE TURNED IN WITH YOUR SAMPLE. THE REQUIRED INFORMATION SHOULD ALSO BE RECORDED IN YOUR NOTEBOOK. THE LABEL ON EACH SAMPLE MUST INCLUDE YOUR NAME AND THE NAME OF THE COMPOUND.

GREEN CHEMISTRY, ATOM ECONOMY, AND ALDOL REACTIONS

Green Chemistry and Sustainability

Chemical industries have contributed to the development of civilized society by introducing numberless valuable products including medicines, fertilizers, pesticides, gasoline, and water purification as well as synthetic materials used in clothing, furniture, and building materials. However, if the earth's entire population were to utilize energy and materials at the same per capita rate as the United States, we would rapidly encounter multiple crises resulting from the depletion of critical resources. A sustainable civilization is one in which we can confidently project achieving a high and equitable standard of living for the earth's population, while at the same time being able to maintain this achievement for the foreseeable future. Highly developed countries have a moral obligation to aid in the advancement of those less developed, and to help build a sustainable civilization. Such an effort contributes to the advancement of society by reducing the pressure on precious natural resources and there by reducing pressure on prices for manufactured goods.

Green Chemistry is the design of chemical products and processes that reduce or eliminate unnecessary materials, waste generation, and hazards in the workplace. Green Chemistry is guided by the twelve principles given below:

1. **Prevention** It is better to prevent waste than to treat or clean up waste after it has been created.

2. **Atom Economy** Synthetic methods should be designed to maximize the incorporation of materials used into the final product.

3. **Less Hazardous Chemical Synthesis** Whenever practicable, synthetic methods should bc dcsigned to use and generate substances that possess little or no toxicity to human health and the environment.

4. **Designing Safer Chemicals** Chemical products should be designed to achieve their desired result while minimizing toxicity.

5. **Safer Solvents and Auxiliaries** The use of auxiliary substances (e.g, solvents, separation agents, etc.) should be made unnecessary wherever possible and innocuous when used.

6. **Design for Energy Efficiency** Energy requirements of chemical processes should be recognized for their environmental and economic impacts and should be minimized. If possible, synthetic methods should be conducted at ambient temperature and pressure.

7. **Use of Renewable Feedstocks** A raw material or feedstock should be renewable rather than depleting whenever technically and economically practicable.

8. **Reduce Derivatives** Unnecessary derivatization (use of blocking groups, protection/deprotection, temporary modification of physical chemical processes) should be minimized or avoided, because such steps require additional reagents and generate waste.

9. **Catalysis** Catalytic reagents (as selective as possible) are superior to stoicheometric reagents.

10. **Design for Degradation** Chemical products should be designed so that at the end of their function they break down into innocuous degradation products and do not persist in the environment.

11. **Real-time analysis for Pollution Prevention** Analytical methodologies should allow for real-time, in-process monitoring and control prior to the formation of hazardous substances.

12. **Inherently Safer Chemistry for Accident Prevention** Substances and the form of a substance used in a chemical process should be chosen to minimize the potential for chemical accidents, including releases, explosions, and fires.

Acknowledgments: The definition of Green Chemistry and the Twelve Principles of Green Chemistry are taken from the American Chemical Society "Introduction to Green Chemistry" website (http//www.chemistry.org/portal/a/c/s/1/acsdisplay.html?DOC=greenchemistryinstitute%5Cwhatare%5C12-principles-green-chemistry.html)

Atom Economy

One way to evaluate the environmental efficiency of a reaction is to determine the amount of the starting materials that ends up in the desired reaction product. Using the balanced equation for a reaction, the percentage atom economy of the reaction is defined as the ratio of the mass of the atoms in the desired product relative to the mass of all the atoms in the starting materials expressed as a percentage.

$$\% \text{ Atom Economy} = 100 \bullet \left(\frac{\text{Formula Weight of All Atoms in Desired Product}}{\text{Formula Weight of All Atoms in Reactants Used}} \right)$$

For example, nitrobenzene can be converted to aniline either using iron metal or molecular hydrogen with a nickel catalyst as the reducing agent. The balanced equations and computations of the percent atom economy for each of these reactions are shown below.

$$4\ C_6H_5NO_2 + 9\ Fe + 4\ H_2O \longrightarrow 4\ C_6H_5NH_2 + 3\ Fe_3O_4$$

Molecular Formulas: $C_{24}H_{20}N_4O_8$ Fe_9 H_8O_4 $C_{24}H_{28}N_4$ Fe_9O_{12}

All Atoms in Reactants | All Atoms in Desired Products

Total Formula Weight: $C_{24}H_{28}N_4O_{12}Fe_9 = 1066.2$ | $C_{24}H_{28}N_4 = 372$

$$\% \text{ Atom Economy} = 100 \cdot \left(\frac{372}{1066.2} \right) = 35\ \%$$

$$C_6H_5NO_2 + 3\ H_2 \xrightarrow[\text{Ni}]{} C_6H_5NH_2 + 2\ H_2O$$

Molecular Formulas: $C_6H_5NO_2$ H_6 C_6H_7N H_4O_2

All Atoms in Reactants | All Atoms in Desired Products

Total Formula Weight: $C_6H_{11}NO_2 = 129$ | $C_6H_7N = 93$

$$\% \text{ Atom Economy} = 100 \cdot \left(\frac{93}{129} \right) = 72\ \%$$

So, based on consideration of percentage atom economy, using molecular hydrogen with a nickel catalyst as the reducing agent would be the preferred method for this synthesis.

The percentage atom economy for a reaction should not be confused with the percent yield for a reaction.

Aldol Reactions

The Aldol reaction is the name given to a condensation reaction that takes place between two carbonyl "partners," one of which contains a hydrogen atom α to its carbonyl group. The reaction takes place by a series of steps, the first of which is the conversion of one partner into the corresponding enol or enolate anion that is catalyzed by acid or base (see the reaction scheme below). Acidic aldol reactions occur through an enol while basic aldol reactions occur via the enolate anion. Both pathways are shown in the reaction scheme below. The enol or enolate anion undergoes nucleophilic addition to the electrophilic carbonyl group of the second partner. In doing so, the nucleophilic carbonyl partner undergoes an α-substitution reaction and the electrophilic carbonyl partner undergoes a nucleophilic addition reaction. The general mechanism is shown in the scheme below.

The initial product of an aldol reaction is a β-hydroxy carbonyl compound (aldol or ketol). If the β-hydroxy carbonyl compound is the desired product from the reaction, the experimental conditions are deliberately chosen to favor isolation of this product from the reaction. The β-hydroxy aldehydes or ketones formed in an aldol reaction can be easily dehydrated to yield α, β-unsaturated products known as conjugated anals or enones respectively. This facile loss of water is how the aldol reaction came to be named as a condensation reaction (ie, water *condenses* from the reaction).

In this laboratory experiment you are performing the base catalyzed aldol condensation (with loss of water) of 3, 4-dimethoxybenzaldehyde with 1-indanone. This reaction was chosen because it demonstrates some of the principles of Green Chemistry. Which of the twelve principles of Green Chemistry are followed in this reaction?

General Mechanism for an Aldol Reaction

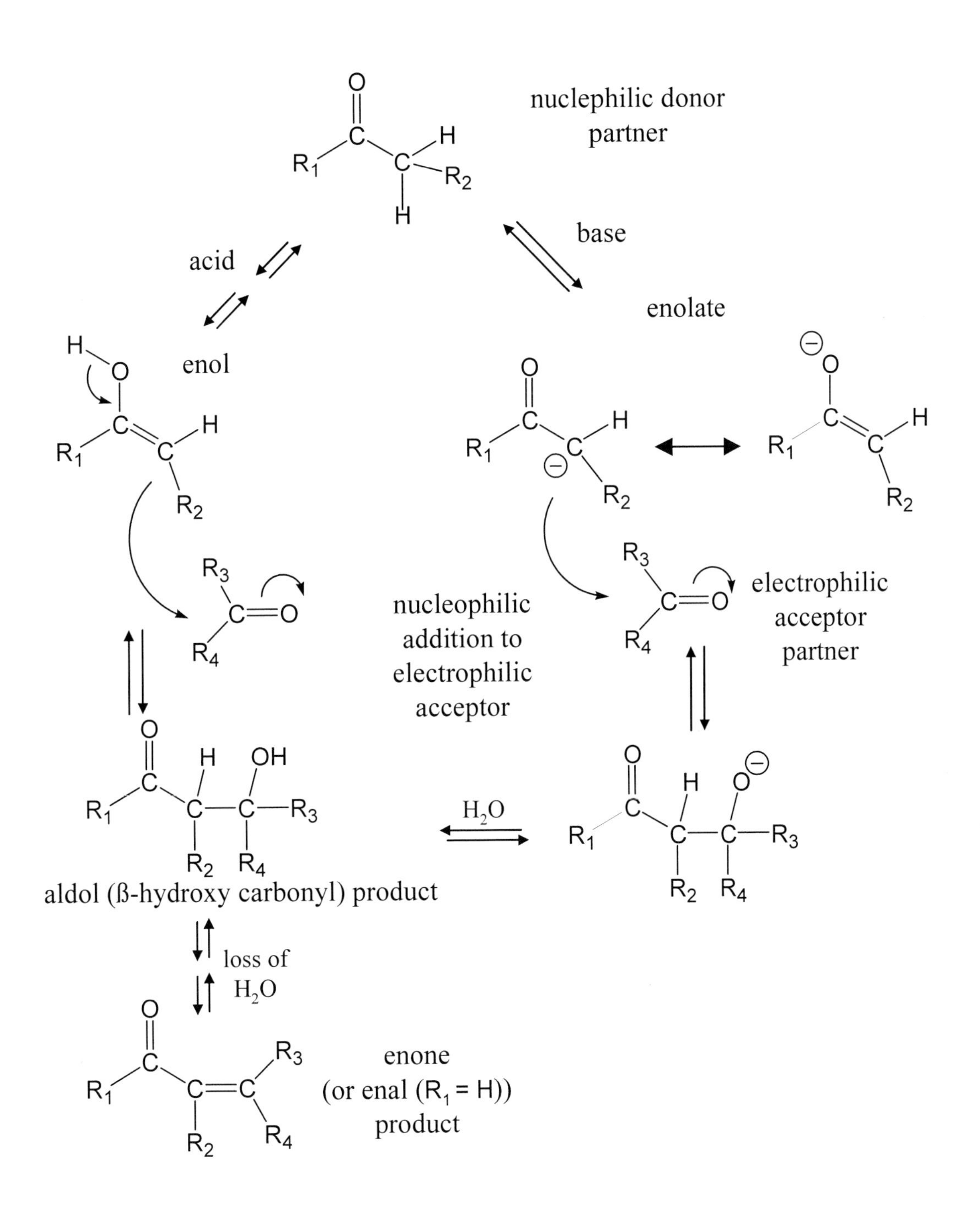

Experiment 8: SOLVENTLESS ALDOL REACTION

CH_3O, CH_3O (3,4-dimethoxybenzaldehyde) + 1-indanone $\xrightarrow{NaOH}$ 2-(3,4-dimethoxybenzylidene)-(indan-1-one) (OCH_3, OCH_3) + H_2O

3,4-Dimethoxy-benzaldehyde	1-Indanone	2-(3,4-Dimethoxyben-zylidene)-(indan-1-one)
MW: 166.17	MW: 132.16	MW: 280.32
mp: 40-43 °C	mp: 38-40 °C	mp: 175-177 °C
.265g	.210g	

Transfer 0.25 g of 3,4-dimethoxybenzaldehyde and 0.20 g of 1-indanone to a 13x100 mm (small test tube). Use your glass stirring rod to mix the solids together until the mixture liquefies. Be careful not to break the test tube. After the mixture has liquified, add 0.05 g of solid NaOH flakes to the test tube and continue mixing until the mixture becomes solid. Once the solid has formed allow the mixture to stand for 15 minutes at room temperature. Then add 2 mL of 10% aqueous HCl solution. You will need to use your stirring rod to suspend the solid in the HCl solution. Use litmus paper to make sure the solution is acidic. If the solution is not acidic add another 2 mL of 10% aqueous HCl solution. Isolate the crude product by vacuum filtration. You may need to use your stirring rod to transfer the pasty solid to the filter paper in the Buchner funnel. Wash the solid with 5 to 10 mL of cold water, then air dry the crude product for 10 minutes, and finally weigh it. Recrystallize all the crude 2-(3,4-dimethoxy-benzylidene)-(indan-1-one) in a 6" test tube using 15 -20 mL of a hot 90 % ethanol/10 % water mixture. Isolate the recrystallized 2-(3,4-dimethoxy-benzylidene)-(indan-1-one) by vacuum filtration, wash it with 10 mL of cold 90 % ethanol/10 % water solution, and then air dry it. Determine the mass and melting point of the recrystallized 2-(3,4-dimethoxy-benzylidene)-(indan-1-one), and also analyze it by GC and IR. After you have completed all the analyses, turn in your recrystallized 2-(3,4-dimethoxy-benzylidene)-(indan-1-one) to your instructor in a labeled test tube. The melting point of the recrystallized 2-(3,4-dimethoxy-benzylidene)-(indan-1-one) should be around 177 °C.

.126g NaOH

0.395g → .195g

Before You Leave: The aqueous filtrate from the isolation of the crude 2-(3,4-dimethoxy-benzylidene)-(indan-1-one) can be flushed down the drain followed by lots of water. The filtrate from the recrystallization should be placed in the organic waste container.

3,4-Dimethoxybenzaldehyde

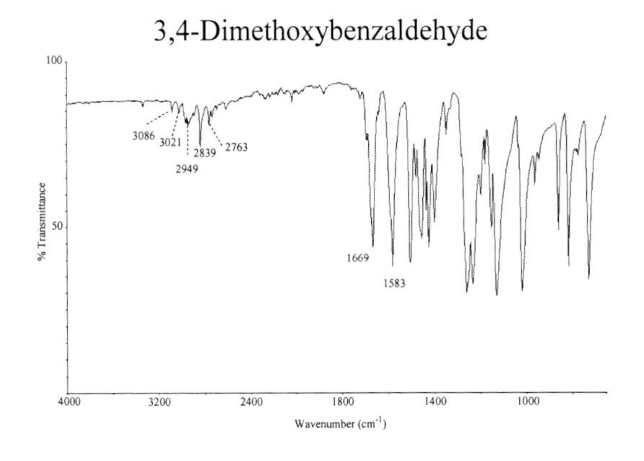

1-Indanone

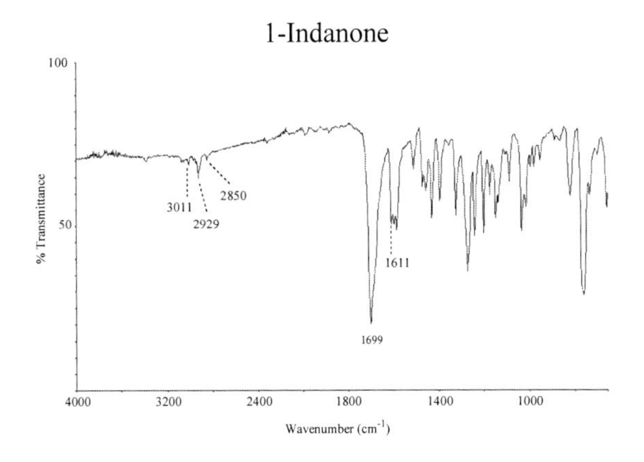

2-(3,4-Dimethoxybenzylidene)-(indan-1-one)

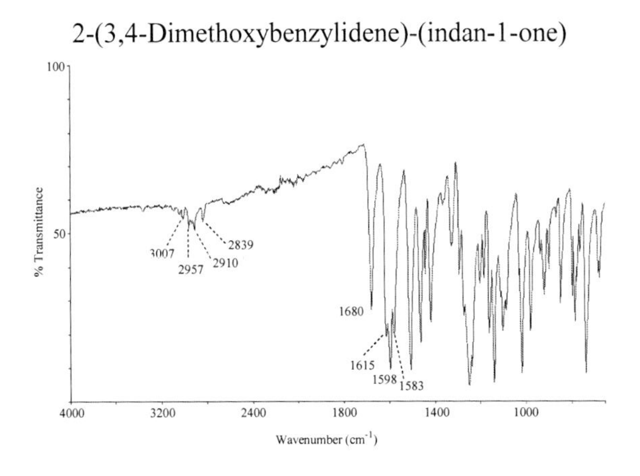

NAME________________________________ I.D.____________

YIELD REPORT EXPERIMENT 8: SOLVENTLESS ALDOL CONDENSATION

WEIGHT OF 3,4-DIMETHOXYBENZALDEHYDE 0.265 g

MOLES OF 3,4-DIMETHOXYBENZALDEHYDE __________ mol

WEIGHT OF 1-INDANONE 0.210 g

MOLES OF 1-INDANONE __________ mol

WEIGHT OF SODIUM HYDROXIDE 0.126 g

MOLES OF SODIUM HYDROXIDE __________ mol

THEORETICAL YIELD OF 2-(3,4-DIMETHOXYBEN-
ZYLIDENE)-(INDAN-1-ONE) __________ g

THEORETICAL YIELD OF 2-(3,4-DIMETHOXYBEN-
ZYLIDENE)-(INDAN-1-ONE) __________ mol

WEIGHT OF CRUDE 2-(3,4-DIMETHOXYBEN-
ZYLIDENE)-(INDAN-1-ONE) OBTAINED __________ g

WEIGHT OF CRUDE 2-(3,4-DIMETHOXYBEN-
ZYLIDENE)-(INDAN-1-ONE) USED IN
RECRYSTALLIZATION __________ g

ACTUAL YIELD OF RECRYSTALLIZED
2-(3,4-DIMETHOXYBENZYLIDENE)-(INDAN-1-ONE) __________ g

PERCENT YIELD OF 2-(3,4-DIMETHOXYBEN-
ZYLIDENE)-(INDAN-1-ONE) __________ %

MELTING POINT (RANGE) OF RECRYSTALLIZED
2-(3,4-DIMETHOXYBEN-ZYLIDENE)-(INDAN-1-ONE) __________ °C

PERCENTAGE ATOM ECONOMY __________ %

USE OF THE BEILSTEIN DATABASE THROUGH MDL CROSSFIRE COMMANDER

The Beilstein database is a structural and factual database in organic chemistry. It currently includes information for over ten million compounds published in the chemical literature from 1771 to the present time. Today the database is online, and can be conveniently accessed through software called CrossFire Commander.

In its most basic form, CrossFire Beilstein can be considered as an enormous handbook containing data such as melting points, boiling points, and densities for a very large number of compounds. It contains literature references (but not always the actual data) to many other kinds of information on these compounds including spectral and pharmacological data as well as preparation and reactions. It can be searched using either chemical structures or text. CrossFire Beilstein can be accessed from any computer on the Howard network which contains the CrossFire Commander software. As of this writing the most recent version of MDL CrossFire Commander is v 7.1.

In the exercises below answer in writing each of the questions shown in **bold** type.

Exercise 1 Looking Up Data on a Specific Compound: Triphenylmethanol (Expt. 7) Using a Structure as the Query

To start the CrossFire program double click the MDL CrossFire Commander 7.1 icon on the desktop (or via the Windows Start Menu - All Programs). The program will open on its Query screen.

Depending upon how the last person exited the program, the message: "No server is connected - please press the Connect button" may appear. If this happens, click on OK, then on Connect (on the row of commands at the top). If you are asked to select a database, check the box for Beilstein, and then click OK.

Click on Draw Structure, which will open up a window for ISIS/DRAW. ISIS/DRAW is an independent chemical structure drawing program that can be used with CrossFire (if you get a different structure drawing program, see Note 1 below). Place the cursor over the various tools along the left side and top of the window to find out what they do. Draw the structure of triphenylmethanol using the "Single Bond" tool, the "Benzene" tool for the phenyl rings, and the "Atom" tool for the oxygen (if the bond tool shows a double or triple bond, hold down the left mouse button on it, and choose the single bond from the resulting menu). All atoms and bonds must be drawn, aside from hydrogens, which may be omitted; shortcuts may not work. When the structure is complete, click the Update and Return to CrossFire Commander tool in the upper left part of the screen. If you make a mistake in ISIS/DRAW, delete it with the Eraser tool or with Lasso Select and the delete key. Clicking on File - New within ISIS/DRAW will make the Update and Return to CrossFire Commander tool disappear.

You should now be back on the Query screen, with the structure of triphenylmethanol displayed in the structure box. Put a bullet in as structure in the Search box. The items in the Free Sites and Allow boxes (top right) should not be checked, and Stereo should be off. Substances should be selected next to "As result I want to get" (bottom of screen). Then click on Start Search.

The resulting screen will show the search progress, and ultimately the number of hits. When the search is complete (very fast in this case), click on View. Notice that a hydrogen atom attached to an oxygen atom is not drawn in the database, i.e., R-OH is shown as R-O. The resulting screen will show pictures of all the hits (just one in this case). Make sure the hit is selected (box around it), and then click on Details or double click on the hit.

The resulting very long page gives the details about triphenylmethanol in the database. Molecular formula and molecular weight are near the top. Scroll down to and look through the Field Availability Lists which give the type of data available for triphenylmethanol. The details of the data follow, beginning with Reaction data. To move quickly to data concerning a particular field of interest, click on the blue code for that field in the Field Availability Lists.

Click on MP to see the data about the melting point of triphenylmethanol. The reported values for the melting point of triphenylmethanol as well as the literature references of these values are given.

1. **What are some of the reported values for the melting point of triphenylmethanol?**

2. **How do these values compare with the mp of your sample of triphenylmethanol?**

Next, move back up to the Field Availability Lists and click on BP and check the data about the boiling point.

3. **What are some of the reported values for the boiling point of triphenylmethanol?**

Now click on NMR in the Field Availability lists. Very few details about the actual spectra are in the database, but literature references are given. It is of course possible to write down a reference, and then look for it in a library. However, if Howard's library subscribes to the online version of the journal in question, you can find and read the article at your computer. You can check for this by clicking on the Full Text box (if an older version of CrossFire Commander is being used, the term LitLink is used).

Find the following NMR reference (at the time of this writing it was 23 of 27 in the NMR section).

> Journal; Rathore, Rajendra; Burns, Carrie L.; Guzei, Ilia A.; JOCEAH; J. Org. Chem.; EN; 69; 5; 2004; 1524 - 1530. (See Note 2 below.)

Click on the Full Text box. You should ultimately see an abstract of this paper. Click on PDF. If Adobe/Acrobat Reader is on your computer, you should see the full text. In many publications, spectra are reported as a series of numbers in a table or experimental section. In this paper, a portion of the actual NMR spectrum of triphenylmethanol is shown. Find the NMR spectrum.

4. What are the approximate chemical shifts shown?

5. What is the solvent?

6. Is the NMR spectrum what you would expect for triphenylmethanol? Briefly justify your answer.

Go back to the CrossFire screen, move up to the Field Availability Lists, and click on PHARM to see pharmacological data for triphenylmethanol.

7. What types of pharmacological effects have been reported?

The Material Safety Data Sheet (MSDS) for a compound also contains information on its health, toxicological, and ecological effects (see sections 3, 11, and 12 of the MSDS). Using a web browser look up the MSDS for triphenylmethanol (see the section in the Introduction on Material Safety Data Sheets for the URL's). The Sigma-Aldrich site also contains physical and spectroscopic data for many of the compounds they sell.

8. What are the potential health effects of triphenylmethanol?

Exercise 2 Looking Up the Preparation of a Specific Compound: 3-Methyl-acetophenone

CrossFire Beilstein is very useful for looking up preparations of compounds. As an example, you will look up methods to prepare 3-methylacetophenone.

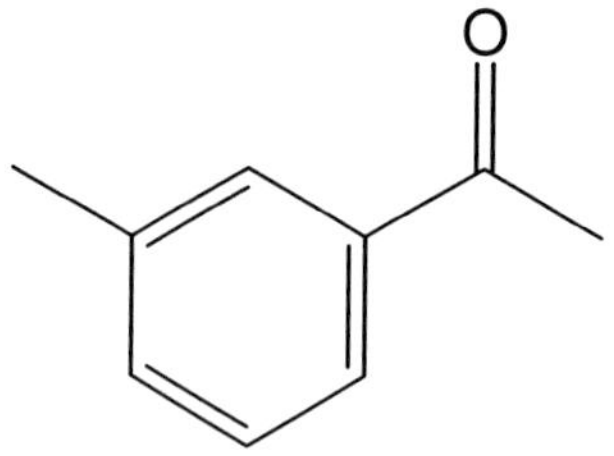

If you are on the Results page, go back to the Query screen by clicking Query at the top of the screen. Then click on Clear Query to prepare for a new search, and then on Draw Structure. Draw the structure of 3-methylacetophenone using ISIS/DRAW, and return to the Query screen as before. If you get a message saying that the object has been changed, and asking if you want to update CrossFire Commander, click yes.

On the Query screen, select as product in the Search Box, and Reactions at the bottom of the page after "As result I want to get". The items in the Free Sites and Allow boxes should not be checked, and Stereo should be off. Click on Start Search.

9. How many hits do you have?

As before, click on View. In this case, the hits will be pictured as reactions (reactants ➜ products). Select the hit which shows the preparation of 3-methylacetophenone from the corresponding alcohol. Click on Details.

10. How many different reactions does the database list for this conversion?

11. Write down one set of reagents and the literature reference.

Click on Grid to get back to the listing of hits. You can look into the preparation of 3-methylacetophenone from other starting materials and find the necessary reagents and references as above.

Exercise 3 Looking Up the Preparation of a Specific Compound From a Specific Starting Material

(A) In experiment 4 you prepared cyclohexanone from cyclohexanol using $Na_2Cr_2O_7$. Assume that you have run out of $Na_2Cr_2O_7$, and that you need to find another reagent for this conversion.

Click on Query to return to the Query screen. Click on Clear Query, then on

Draw Structure, and then in ISIS/DRAW draw cyclohexanol on the left side of the screen and cyclohexanone on the right side of the screen. Using the arrow tool, draw an arrow from cyclohexanol to cyclohexanone. Click the Update and Return to CrossFire Commander button to get to the Query screen as usual. The only option in the Search box will be as reaction. The items in the Free Sites and Allow boxes should not be checked, and the Stereo button should be off. Reactions should be selected under "As result I want to get". Perform the search (Start Search). You will get a large number of hits.

12. How many hits did you get?

13. Write down the details as to reagents and literature reference for one of the hits in which a halogen containing oxidizing agent was used.

14. What was the reported percentage yield? (You will need to look up the original paper, so pick a hit that you can look up on line.)

(B) In experiment 6 you prepared methyl 3-nitrobenzoate from methyl benzoate using HNO_3 and H_2SO_4. The reported isomer distribution is given in Table 1 of this experiment.

What other reagents/conditions can you find using CrossFire Beilstein to do this conversion? See Note 3 below regarding drawing nitro groups in CrossFire Beilstein. To find another reagent for which the percentages of the three methyl nitrobenzoate isomers are reported, make sure you select a hit in which all three isomers are shown. You can find the percentages by looking up the actual papers. However, most are listed in the database. The compounds are listed by their Beilstein Registry numbers (BRN). The reactant and product BRN's are given at the top of the Results/Details screen.

15. What is another reagent for this conversion in which the percentages of the three product isomers are reported?

16. What is the literature reference?

17. What are the percentages for each product isomer?

18. How do these percentages compare to the percentages obtained with HNO_3 and H_2SO_4?

Notes:

1. Other structure drawing programs are also available and can be used. The last one used will come up when you click Draw Structure. You can choose ISIS/DRAW from the Query screen (Options - Select Structure Editor).

2. This format indicates a paper by R. Rathore, C. L. Burns, and I. A. Guzei, in the Journal of Organic Chemistry, Volume 69, issue number 5, pages 1524-1530, published in 2004. The "EN" indicates the paper is written in English. If this paper is cited as a literature reference in an American Chemical Society journal, the appropriate format is:

   Rathore, R.; Burns, C. L.; Guzei, I. A. *J. Org. Chem.* **2004**, *69*, 1524-1530.

3. In the Beilstein database, the nitro group is drawn with five bonds to the nitrogen.

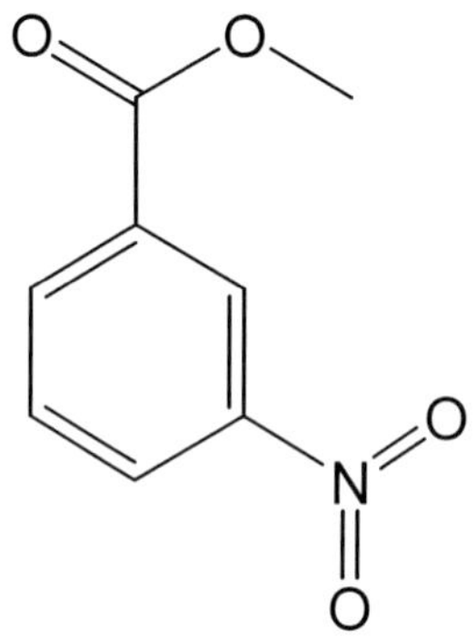

For example methyl 3-nitrobenzoate is drawn as

   You may not get any hits if you draw the nitro group differently. However, do not draw nitro groups in this manner for your organic chemistry classes: You will get into trouble and lose points.

4. The Beilstein database was originally published in German. Most of it has been translated into English, but occasionally you may encounter some German.

IDENTIFICATION OF AN UNKNOWN ORGANIC COMPOUND

You will receive a liquid and/or a solid organic compound, whose identity you will determine by chemical and spectroscopic methods. About four lab periods are devoted to this unit. The procedures you use will depend on the nature of your unknown compound.

CAUTION: All Unknown Compounds are to be considered toxic! Do Not Smell, Taste, or Touch them. Clean up all spills immediately

Each unknown compound will contain *at least* one or more of the following functional groups: carboxylic acid, alcohol, aldehyde, ketone, ester, amine, amide, nitrile, phenol. In addition, the compound may be aromatic, olefinic, or aliphatic, and it may contain nitro groups, ethers, and halogens.

The following outline is a guide to obtaining the identity of your compound in an efficient manner. Directions for the physical and chemical tests appear later in this section. Record the results of your tests in your notebook; use a separate page for each unknown.

1. Physical state. The unknowns are commercial samples and their purity varies. However, you should be able to obtain a melting point value close to that reported for the pure compound. Distillation and recrystallization is generally unnecessary. The melting point of a solid should be determined. The melting point is determined in the usual way. You will be given the approximate bp of any liquid unknown. The value you are given will be somewhere in the range: Known BP ± 10 °C.

The color and odor (CAUTION: CAUTIOUSLY SMELL THE CORK, NOT THE COMPOUND) should be noted, and the Combustion test and Beilstein test should be carried out. The Beilstein test is an easy, sensitive test for the presence of halogen. You should run the test on a compound which is known to contain halogen and one which does not, and compare the results with the test on your unknown. Comparing results for unknowns to known, reference compounds, is a standard laboratory practice. A list of available reference compounds is found just after the section on Chemical Tests (Color Reactions).

2. Infrared spectrum. Tables for interpretation of the IR spectrum and spectra of several compounds appear earlier in this manual. Other information may be found in your organic lecture textbook. A flow chart and summary sheet for analyzing the IR spectrum of an unknown compound are included after this section. Label each spectrum

with your name, sample name (e.g.,"unknown liquid"), the date, and phase (liquid film, pure solid, Nujol mull, or evaporate).

3. Solubility tests. Use 5-6 drops of liquid or 1-2 mg of solid unknown in 2-3 mL of solvent to determine solubility. Follow the "Solubility Tests Flow Chart" for the solubility tests. Compounds which dissolve in each solvent are listed.

> Water. Very polar, polyfunctional and/or low molecular weight compounds which contain various functional groups, e.g., acetic acid, methylamine, methanol, acetone, sucrose. If your unknown dissolves in water, test the solution with litmus or pH paper to see if it is acidic (carboxylic acid) or basic (amine). If the unknown is soluble in water, do not test solubility in the following aqueous solutions.
>
> 5% HCl (~1.5 M). Only amines will be soluble. Aromatic amines containing nitro or other electron-withdrawing groups may be less soluble.
>
> 5-10% NaOH (~1.5 M). Carboxylic acids and phenols. If the compound is soluble, then try $NaHCO_3$; if it is insoluble, skip $NaHCO_3$.
>
> 5-10% $NaHCO_3$ (~0.6 M). Carboxylic acids and a few phenols which contain nitro groups.
>
> Concentrated Sulfuric Acid. Organic compounds containing oxygen or nitrogen that are not soluble in any of the above reagents (aldehydes, ketones, esters, amides, ethers, alcohols) or alkenes and alkynes react with sulfuric acid and so appear to be soluble in it.

4. IR Spectroscopy & Solubility tests. Analysis of the infrared spectrum and the solubility tests will allow assignment to probable functional groups. In the table below, [+] means the compound shows absorption in the particular IR region or is soluble, [-] means the compound shows no absorption in the particular IR region or is insoluble. Some unknown compounds may contain more than one functional group, e.g., phenol and ester, so use this table carefully.

5. Chemical Tests. Aldehydes and ketones react with 2,4-dinitrophenylhydrazine (2,4-DNP) reagent; esters and amides do not react. Aldehydes give a positive test with Tollens' reagent and ketones do not. Primary and secondary alcohols and aldehydes give a positive test with chromic acid reagent (Jones oxidation) and tertiary alcohols do not. Directions for these and other chemical tests are given in later in this section.

$1700\ cm^{-1}$	$3400\ cm^{-1}$	NaOH	HCl	Possible Functional Groups
+	-	-	-	aldehyde, ketone, ester, tertiary amide
+	+	-	-	hydroxy-aldehyde, hydroxy-ketone, hydroxy-ester; primary or secondary amide
+	+	+	-	carboxylic acid (also soluble in $NaHCO_3$)
-	+	+	-	phenol
-	+	-	+	primary or secondary amine
-	+	-	-	alcohol
-	-	-	+	tertiary amine
-	-	-	-	nitrile (IR $2200\ cm^{-1}$)

6. Molecular weight and ^{1}H-NMR spectrum. You will turn-in a preliminary report for your unknown which will contain the mp or bp, results of the solubility, flame, IR, and chemical tests and your conclusions about the functional group(s) present in the compound. If you have correctly determined the mp or bp (within 5°C) and correctly identified the functional group(s), you will receive a copy of the ^{1}H- NMR spectrum and the mass spectrum of your unknown compound. You will use the mass spectrum to figure out the molecular weight of your unknown compound. Your final report must show the identity of the compound, and an interpretation of the ^{1}H-NMR spectrum, which you will return with your report.

Tables of compounds arranged by functional group and bp/mp follow the descriptions of the chemical tests. Additional tables will be available.

FLAME TESTS

The way an organic compound burns can give information about its structure. You will perform two different types of flame test on your unknowns. A Combustion test (or ignition test) can indicate that the compound contains an aromatic ring. A Beilstein test can indicate that the compound contains halogen (chlorine, bromine and/or iodine). In a Combustion test you observe the type of smoke given off when a compound burns. Aromatic compounds (i.e., organic compounds that have a high percentage of carbon) produce a sooty, black smoke when they burn. In a Beilstein test you observe the color of the flame when a compound is burned on clean copper wire. Compounds containing halogen (chlorine, bromine, or iodine) burn with an intense green flame when they are burned on copper wire. The Beilstein test is very sensitive: fingerprints (containing sodium chloride) may give a positive result.

Combustion: Clean your metal spatula by holding it in the flame of a Bunsen burner for 20 to 30 seconds. Allow the spatula to cool to room temperature, making sure not to touch the metal part of the spatula with your fingers. After the spatula has completely cooled, place a few drops of your liquid sample or a small pinch of your solid sample on the end of the metal spatula. Place the end of the spatula with your sample on it in the flame of the Bunsen burner until the sample ignites, then remove the spatula from the flame of the Bunsen burner. Observe the combustion of your sample. Does it burn with sooty, black smoke, or does it burn without giving off any smoke? Does it leave any residue on the spatula? (If there is a residue, allow the spatula to cool and then add a drop of water to the residue and then test with pH paper.) Record your observations in your notebook. Try combusting samples of benzoic acid (solid) and toluene (liquid). Since these compounds are aromatic, each should burn giving off sooty, black smoke. Try combusting samples of crotonic acid (solid) and cyclohexane (liquid). Since these compounds are aliphatic, each should burn without giving off any smoke. Then try combusting a sample of each of your unknowns. Be sure to clean your spatula between each sample.

Beilstein: Obtain a length of copper wire and a cork from the stockroom. Bend one end of the copper wire into a tight loop. Insert the other end of the copper wire into the cork, so that you can use the cork to hold the copper wire without touching the wire itself. Clean the loop end of the copper wire by inserting it in the flame of a Bunsen burner for 10 to 15 seconds (heating the loop end red hot is good enough). Allow the copper wire to cool making sure you do not touch it or let it touch the lab bench. After the wire has completely cooled place a few drops of a liquid sample or a small pinch of a solid sample on the loop end of the copper wire. Place the loop end of the copper wire in the flame of a Bunsen burner to ignite the sample. Hold the wire in the flame until the wire begins to glow. An intense green flash indicates that the sample contains halogen (chlorine, bromine, and/or iodine). Record your observations in your notebook. Try this test on a sample of 4-chloroacetanalide (solid) and bromobenzene (liquid). Since these compounds contain halogen, each should produce an intense green flash. Try this test on sample compounds that do not contain halogen. Repeat this test using a sample of each of your unknowns. Make sure you have cleaned the copper loop between each sample.

CHEMICAL TESTS (COLOR REACTIONS)

Your instructor will tell you which of the following chemical tests must be performed on your unknown compound(s).

2,4-Dinitrophenylhydrazine (2,4-DNP):

Aldehydes and *ketones* react with 2,4-dinitrophenylhydrazine to give 2,4-dinitrophenylhydrazones, which are usually red, orange, or yellow solids.

O, C, R, R_1 + O_2N, NH—NH_2, NO_2 → O_2N, NH—N=C, R, R_1, NO_2

aldehyde or ketone

2,4-dinitrophenylhydrazine

2,4-dinitrophenylhydrazone of the aldehyde or ketone

Procedure: In a clean dry small test tube (remember acetone itself will give a positive reaction in this test) place 2 drops of liquid sample or 100 mg of solid sample. Add about 2 mL of 95% Ethanol to dissolve the sample. Add 2 mL of the 2,4-dinitrophenylhydrazine reagent and mix well. If a red, orange, or yellow precipitate does not form immediately, allow the mixture to stand for 15 minutes. Record your observations in your notebook.

Reference compounds that will react in this test are: benzaldehyde, acetone, acetophenone, or benzophenone.

Iodoform Reaction (I_2/NaOH):

Methyl ketones, *acetaldehyde* and *secondary alcohols having a -CHOHCH*$_3$ group react with I_2 in basic solution to produce iodoform (CHI_3) which is a yellow solid.

$$R-C(=O)-CH_3 \text{ (methyl ketone)} \quad \text{or} \quad R-CH(OH)-CH_3 \text{ (hydroxymethyl secondary alcohol)} \xrightarrow[NaOH]{I_2} R-C(=O)-O^- Na^+ \text{ (carboxylate salt)} + CHI_3 \text{ (Iodoform, yellow ppt)}$$

Procedure: a. Sample compound is soluble in water. In a small clean test tube (remember acetone, ethanol, and isopropanol will give positive reactions in this test) place 2 drops of liquid sample or 50 mg of solid sample. Add 2 mL of water to dissolve the sample. Then add 2 mL of 3 M NaOH and mix. Now slowly add 3 mL of Iodine/KI solution with shaking. The brown/red color of Iodine will disappear and a yellow layer will separate if the sample compound gives a positive reaction.

b. Sample compound is not soluble in water. In an 8" clean test tube (remember acetone , ethanol, and isopropanol will give positive reactions in this test) place 2 drops of liquid sample or 50 mg of solid sample. Dissolve the sample compound in 2 mL of dioxane. Add 2 mL of water and mix. Then add 2 mL of 3 M NaOH and mix thoroughly. Now slowly add 3 mL of Iodine/KI solution with shaking. After all the Iodine/KI solution has been added, add 10 mL of water and mix completely. The brown/red color of Iodine will disappear and a yellow layer will separate if the sample compound gives a positive reaction.

Acetone or acetophenone will produce iodoform in this reaction.

Tollens' Reaction ($Ag(NH_3)_2OH$):

Aldehydes can be distinguished from ketones by the Tollens' reaction. An aldehyde is easily oxidized to the corresponding carboxylic acid by mild oxidizing agents such as ammoniacal silver hydroxide ($Ag(NH_3)_2OH$). In this process the silver ion is reduced to silver metal, which can be deposited as a silver mirror on the clean walls of the reaction tube. The actual reagent used in this test must be made up just before the reaction is carried out.

$$\underset{\text{aldehyde}}{RC(=O)H} + \underset{\text{Tollens' reagent}}{Ag(NH_3)_2OH} \longrightarrow \underset{\text{Silver metal}}{Ag} + \underset{\text{carboxylate salt}}{RC(=O)O^- \; NH_4^+}$$

Procedure: Place 3 mL of Tollens' Solution A (0.3 M $AgNO_3$) into a clean, dry small test tube. Add concentrated ammonium hydroxide (NH_4OH) dropwise with shaking until the brown/grey precipitate just dissolves. The solution may be slightly grayish, but it should be reasonably clear. Now add 3 mL of Tollens Solution B (1.3 M KOH) to the test tube containing the clear, grayish solution. Again add concentrated ammonium hydroxide dropwise with shaking until the solution is almost clear. This solution is the Tollens' Reagent.

Using a clean pasteur pipette add 20 drops of the Tollens' reagent you just prepared to a clean, dry small test tube. For a liquid sample, add 1 drop of the liquid to the Tollens' reagent using a clean pasteur pipette and mix well. For a solid sample add a few crystals to the Tollens' reagent and mix well. Formation of a silver mirror or a black precipitate is a positive reaction. If no reaction occurs at room temperature, heat the reaction mixture for a few minutes in a warm water bath.

Benzaldehyde will react to produce a silver mirror in this test.

Br_2/CH_2Cl_2

Most *alkenes* and *alkynes* react rapidly with bromine (Br_2) dissolved in dichloromethane (CH_2Cl_2) to give alkyl bromides. Bromine dissolved in dichloromethane (Br_2/CH_2Cl_2) has a red/brown color. Alkenes, alkynes, and alkyl bromides are colorless.

$$R_1R_2C{=}CR_3R_4 \xrightarrow[CCl_4]{Br_2\ \text{(red-brown)}} R_1R_2CBr{-}CBrR_3R_4$$

alkene → alkyl dibromide

$$R_1{-}C{\equiv}C{-}R_2 \xrightarrow[CCl_4]{Br_2\ \text{(red-brown)}} R_1{-}CBr_2{-}CBr_2{-}R_2$$

alkyne → alkyl tetrabromide

Procedure: Place 2 drops of a liquid sample in a clean, dry small test tube. For a solid sample dissolve a few crystals of the solid in 2 drops of diethyl ether. Add the bromine in dichloromethane solution one drop at a time to the sample solution, shaking the sample solution after each addition. Observe whether the color of the bromine in dichloromethane solution disappeared or was just diluted. Count the number of drops of bromine in dichloromethane solution you need to add for the red/brown color to persist. If more than 5 drops is required the sample compound is an alkene or an alkyne.

Cyclohexene will react in this test. Compare the reaction of cyclohexene to that of cyclohexane or toluene. Neither toluene or cyclohexane will react in this test.

$KMnO_4$/NaOH (Bayer Test):

Alkenes and *alkynes* also react with (are oxidized by) potassium permanagnate ($KMnO_4$, purple color) in basic solution (NaOH) to give vicinal diols or carboxylic acids, respectively, and MnO_2 (a brown precipitate). The alkenes, alkynes, diols, and carboxylic acids are all colorless. Other easily oxidized organic compounds such as aldehydes and amines may also react in this test.

$$R_1R_2C{=}CR_3R_4 \xrightarrow[NaOH]{KMnO_4 \text{ (purple)}} R_1R_2C(OH){-}C(OH)R_3R_4 + MnO_2$$

alkene → vicinal diol + manganese dioxide (brown ppt)

$$R_1{-}C{\equiv}C{-}R_2 \xrightarrow[NaOH]{KMnO_4 \text{ (purple)}} R_1COO^- + R_2COO^- + MnO_2$$

alkyne → carboxylic acid + carboxylic acid + manganese dioxide (brown ppt)

Procedure: In a clean dry test tube, dissolve 1 drop of a liquid sample compound or a few crystals of a solid sample compound in 2 mL of acetone. Add $KMnO_4$/NaOH solution one drop at a time to the sample solution, shaking the sample solution after each addition. Observe whether the purple color of the $KMnO_4$/NaOH solution disappeared or was just diluted. A black or brown precipitate of MnO_2 will form if a reaction occurs. Count the number of drops of $KMnO_4$/NaOH solution you need to add for the purple color to persist. If more than a few drops is required the sample compound is an alkene or an alkyne.

Cyclohexene will react in this test. Compare the reaction of cyclohexene to that of cyclohexane or toluene. Neither toluene nor cyclohexane will react in this test.

Br_2/H_2O:

Phenols and *other aromatic rings having very activating substituents* will react with bromine in water in an electrophillic substitution reaction in which bromine is substituted at all unsubstituted positions ortho or para to the phenolic OH group.

$$\text{phenol} \xrightarrow[H_2O]{Br_2\ \text{(red-brown)}} \text{2, 4, 6-tribromophenol}$$

Procedure: To a clean 6" test tube add 2 drops of a liquid sample or 0.1 g of a solid sample and 5 mL of water. If the sample does not dissolve in the water add just enough 3 M NaOH solution to dissolve the sample. Then add dropwise the red/brown Br_2/H_2O solution. After addition of each drop of Br_2/H_2O reagent shake the solution and observe whether the red/brown color of the Br_2/H_2O solution disappeared or was just diluted. Count the number of drops of Br_2/H_2O solution you need to add for the red/brown color to persist. A precipitate (the brominated substitution product) may also form as the color disappears.
Phenol and aniline will react in this test.

H_2CrO_4/H_2SO_4 (Jones Reagent):

Primary and secondary alcohols as well as *aldehydes* are easily oxidized by oxidizing agents such as H_2CrO_4/H_2SO_4 (Jones Reagent). In this reaction the orange/ brown Cr(VI) of the chromic acid (H_2CrO_4) is reduced to Cr(III) which is green. Tertiary alcohols are not easily oxidized.

Note this reagent was used to prepare cyclohexanone from cyclohexanol, a secondary alcohol, in experiment 4.

$$R{-}CH_2OH \text{ (primary alcohol)} \quad \text{or} \quad R{-}CHO \text{ (aldehyde)} \xrightarrow[\text{(red/brown)}]{H_2CrO_4} R{-}COOH \text{ (carboxylic acid)} + Cr^{+3} \text{ (green)}$$

$$R_1{-}CH(OH){-}R_2 \text{ (secondary alcohol)} \xrightarrow[\text{(red/brown)}]{H_2CrO_4} R_1{-}CO{-}R_2 \text{ (ketone)} + Cr^{+3} \text{ (green)}$$

$$R_1{-}C(OH)(R_2){-}R_3 \text{ (tertiary alcohol)} \xrightarrow[\text{(red/brown)}]{H_2CrO_4} \text{No Reaction}$$

Procedure: In a small, dry test tube dissolve 1 drop of a liquid sample or a few crystals of a solid sample in 1 mL of acetone. Add 1 drop of the H_2CrO_4/H_2SO_4 (Jones Reagent) solution and mix well by shaking. Observe any change in color that occurs in less than 1 minute. A change in color from orange/brown to green indicates a positive reaction. 1-Butanol, 2-butanol, benzyl alcohol, ethanol, and benzaldehyde will react in this test. t-Butanol, a tertiary alcohol, will not react with Jones reagent.

$ZnCl_2$/HCl (Lucas Test):

Alcohols (*secondary*, *tertiary*, and *allyl*) that can produce stable carbocations react rapidly with concentrated HCl containing $ZnCl_2$ (Lucas reagent) to produce alkyl chlorides that are insoluble in or immiscible with the reaction solution. Tertiary alcohols and allylic alcohols react very rapidly with Lucas reagent at room temperature while secondary alcohols may take up to 5 minutes to react at room temperature. Primary alcohols do not react at room temperature.

$$R-CH_2OH \xrightarrow[HCl]{ZnCl_2} \text{No Reaction}$$

primary alcohol

$$R_1-CH(OH)-R_2 \xrightarrow[HCl]{ZnCl_2} R_1-CH(Cl)-R_2$$

secondary alcohol → secondary alkyl chloride (ppt)

$$R_1-C(OH)(R_2)-R_3 \xrightarrow[HCl]{ZnCl_2} R_1-C(Cl)(R_2)-R_3$$

tertiary alcohol → tertiary alkyl chloride (ppt)

Procedure: To a clean, dry small test tube add 3 mL of Lucas reagent ($ZnCl_2$/HCl). Add 2 drops of a liquid alcohol or a few crystals of a solid alcohol to the test tube, stopper the tube and then shake it. Look for the formation of a cloudy emulsion or a second layer in the test tube and note the time required for this reaction to occur (essentially immediate, 5-10 minutes required, or no reaction). In order for this test to work the alcohol must be soluble in the Lucas reagent.

t-Butanol or benzyl alcohol will react essentially immediately, while 2-butanol may take a few minutes to react. 1-Butanol will not react.

$AgNO_3$/EtOH:

Alkyl halides but not aryl halides will react with silver nitrate solution to form an insoluble silver halide precipitate.

$$\underset{\text{alkyl halide}}{R{-}X} \xrightarrow[\text{EtOH}]{AgNO_3} \underset{\text{silver halide (ppt)}}{AgX} + \underset{\text{alkyl nitrate}}{R{-}ONO_2}$$

Procedure: To a small, dry test tube add 1 drop of a liquid sample or a few crystals of a solid sample. If a solid sample was used add 5 drops of ethanol (EtOH) to dissolve the solid. Add 1 mL of $AgNO_3$/EtOH (silver nitrate in ethanol) solution and mix by shaking. If no precipitate forms after 5 minutes, heat the reaction mixture on a steam bath for 3-4 minutes. If any precipitate formed either before or after heating, add 2 drops of 1 M nitric acid to the reaction mixture containing the precipitate, shake, and observe if the precipitate dissolves. A precipitate that does not dissolve in nitric acid indicates that the sample contained an alkyl halide.

Benzyl bromide, but not bromobenzene, reacts in this test.

Hinsberg Test:

Primary, *secondary*, and *tertiary amines* can be distinguished by their reactions with benzenesulfonyl chloride in basic aqueous solution. Both primary and secondary amines react with benzenesulfonyl chloride in basic aqueous solution to give the corresponding benzenesulfonamide, which is insoluble in water or acidic solution. Tertiary amines do not react. The benzenesulfonamide produced by a primary amine behaves differently from the benzenesulfonamide produced by a secondary amine in basic aqueous solution due to the different number of hydrogen atoms present on the amide nitrogen of each benzenesulfonamide (a sulfonamide formed from a primary amine will have 1 amide hydrogen while a sulfonamide formed from a secondary amine will not have any amide hydrogens). Since the amide hydrogen of a sulfonamide is relatively acidic, the benzenesulfonamide produced from a primary amine can be deprotonated in basic aqueous solution to give a soluble salt, where as the benzenesulfonamide formed from a secondary amine can not be deprotonated and so is insoluble in basic aqueous solution.

$$R{-}\ddot{N}H_2 + C_6H_5{-}SO_2{-}Cl \xrightarrow{KOH} C_6H_5{-}SO_2{-}\ddot{N}^{\ominus}{-}R\ K^{\oplus} \xrightarrow{HCl} C_6H_5{-}SO_2{-}NH{-}R$$

primary amine + benzenesulfonyl chloride → potassium benzenesulfonamide (soluble) → benzenesulfonamide (insoluble)

$$R_1{-}\ddot{N}H{-}R_2 + C_6H_5{-}SO_2{-}Cl \xrightarrow{KOH} C_6H_5{-}SO_2{-}\ddot{N}(R_2){-}R_1 \xrightarrow{HCl} \text{No Reaction}$$

secondary amine + benzenesulfonyl chloride → benzenesulfonamide (insoluble)

$$R_1{-}\ddot{N}(R_2){-}R_3 + C_6H_5{-}SO_2{-}Cl \xrightarrow{KOH} \text{No Reaction} \xrightarrow{HCl} \text{No Reaction}$$

tertiary amine + benzenesulfonyl chloride

Procedure: To a clean 6" test tube add 5 drops of a liquid amine sample or 0.1 g of a solid amine sample followed by 5 mL of 2 M potassium hydroxide (KOH) solution. Then add 10 drops of benzenesulfonyl chloride, stopper the test tube and shake it intermittently for 3-5 minutes. Using pH paper check to see that the reaction mixture is strongly basic. If the reaction mixture is not strongly basic add more 2 M potassium hydroxide (KOH) solution dropwise with shaking until the reaction mixture is strongly basic. Observe whether the strongly basic solution is homogenous, whether a precipitate, or whether two layers are present.

If two layers are present, remove some of the upper, organic layer. Add 3 mL of 5% HCl solution to the upper, organic layer. Using litmus paper verify that the aqueous layer is still acidic (add more 5% HCl solution if necessary). If the upper organic layer is soluble in 5% HCl solution the sample amine was a tertiary amine. If the upper organic layer is not soluble in 5% HCl solution, the sample amine was a secondary amine.

If a precipitate is present in the strongly basic solution, add 5%

HCl solution to the mixture unitl the solution is acidic to litmus paper. If the precipitate does not dissolve, then the sample amine was a secondary amine.

If the strongly basic solution is homogenous add 5% HCl dropwise to it with shaking after each addition until the solution is just acidic to litmus paper. Did a precipitate form when the solution was acidified? If a precipitate formed when the solution was acidified then the sample amine was a primary amine. If no precipitate formed when the solution was acidified, then the sample amine was a tertiary amine.

Reference Compounds for Flame and Chemical Tests

Alcohol	primary	1-butanol (n-butanol)
	secondary	2-butanol
	tertiary	tert-butanol (2-methyl-2-propanol)
Aldehyde		benzaldehyde
Alkene		cyclohexene
Amide		4-chloroacetanalide[B]
Amine	primary	n-hexylamine
	secondary	di-n-propylamine
	tertiary	triethylamine
Carboxylic acid		benzoic acid [C]
		crotonic acid [C]
Ester		methyl benzoate
Ether		diethyl ether
Halide	alkyl	benzyl bromide [B]
	aryl	bromobenzene [B]
Hydrocarbon	aliphatic	cyclohexane [C]
	aromatic	toluene [C]
Ketone		acetone
		acetophenone
		benzophenone
Phenol		phenol

[B]use in Beilstein flame test
[C]use in Combustion flame test

Solubility Tests Flow Chart

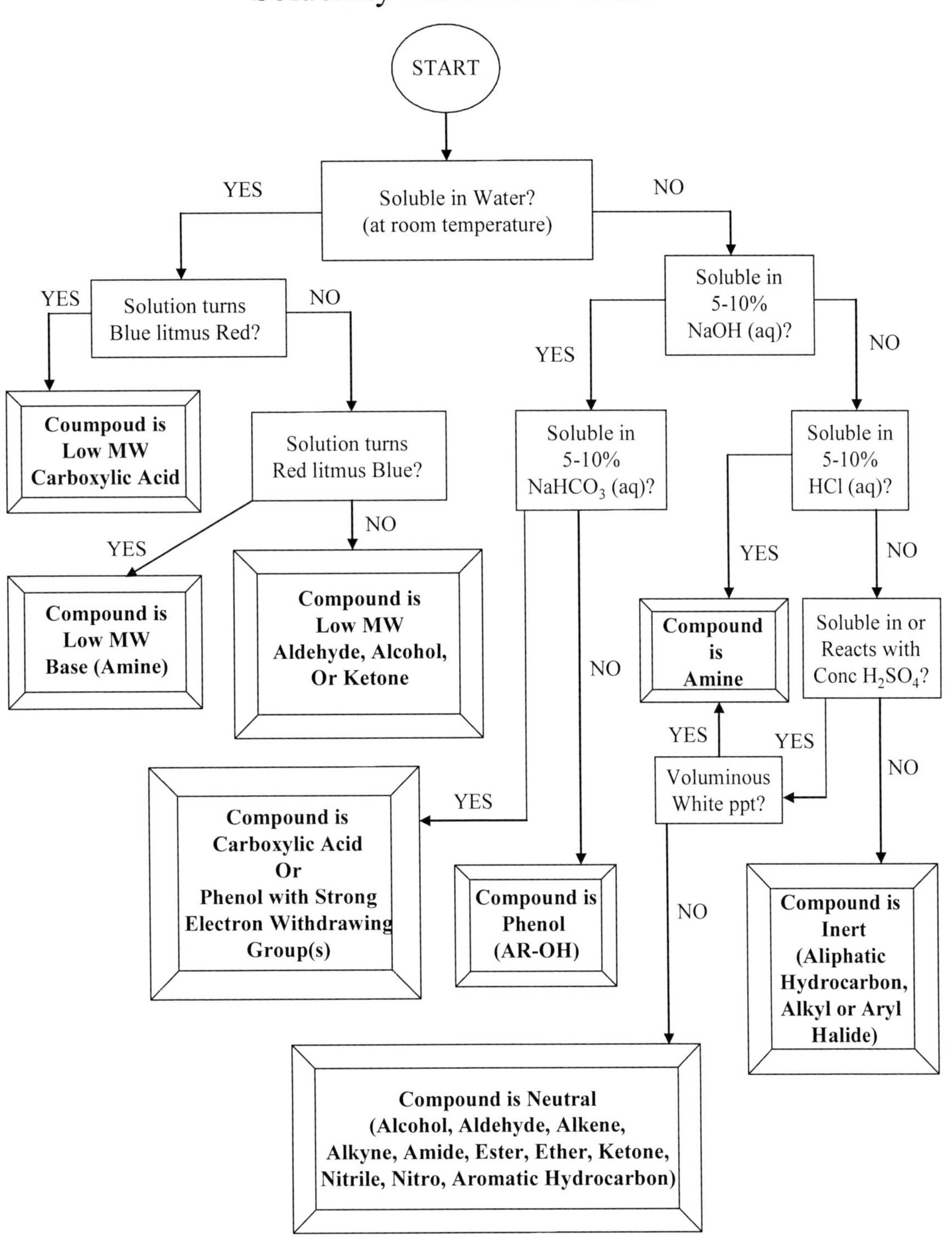

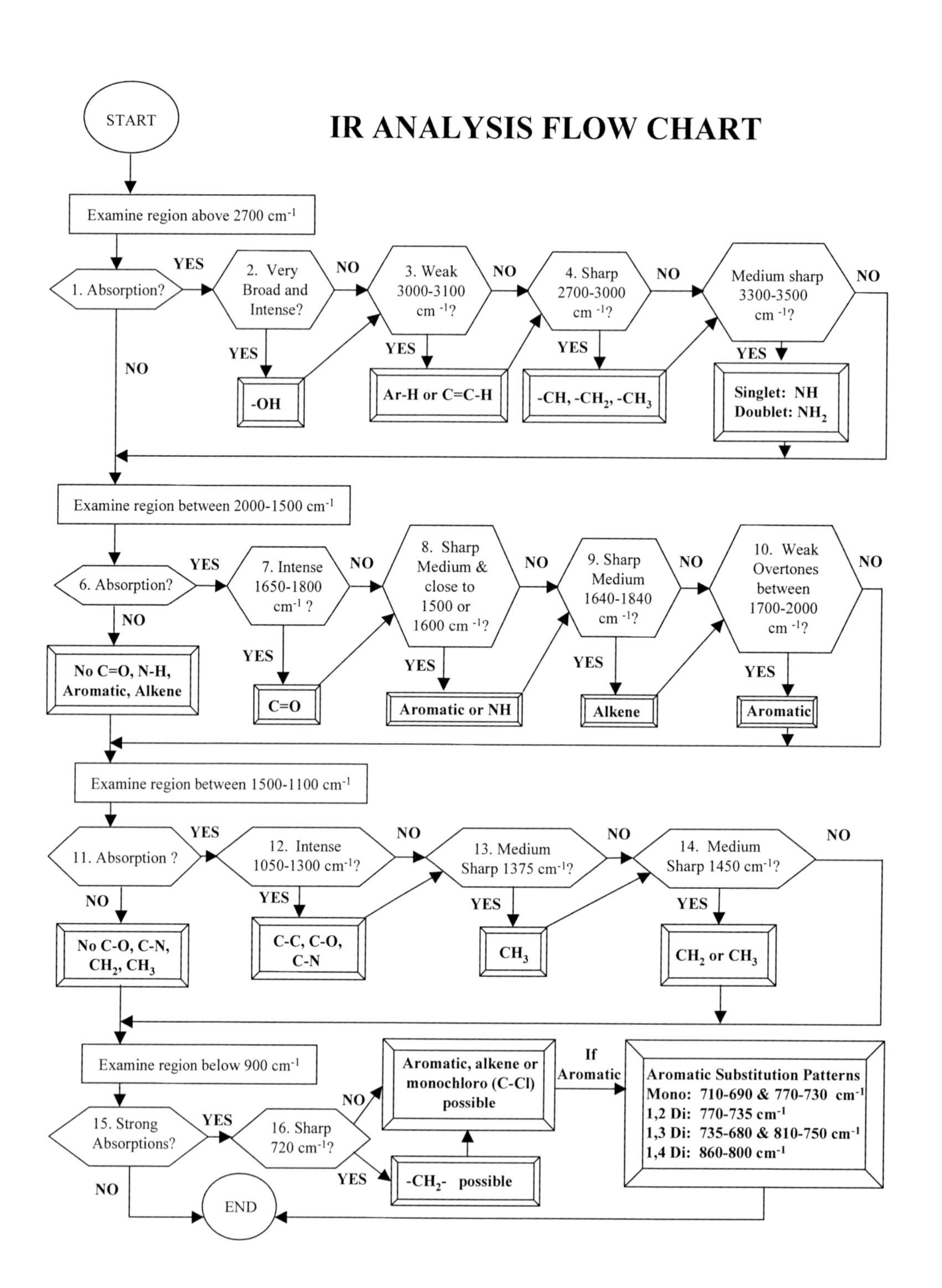
START
IR ANALYSIS FLOW CHART
Examine region above 2700 cm-1
1. Absorption?
YES
2. Very Broad and Intense?
NO
3. Weak 3000-3100 cm -1?
NO
4. Sharp 2700-3000 cm -1?
NO
Medium sharp 3300-3500 cm -1?
NO
YES
-OH
YES
Ar-H or C=C-H
YES
-CH, -CH2, -CH3
YES
Singlet: NH
Doublet: NH2
NO
Examine region between 2000-1500 cm-1
6. Absorption?
YES
7. Intense 1650-1800 cm-1 ?
NO
8. Sharp Medium & close to 1500 or 1600 cm -1?
NO
9. Sharp Medium 1640-1840 cm -1?
NO
10. Weak Overtones between 1700-2000 cm -1?
NO
NO
No C=O, N-H, Aromatic, Alkene
YES
C=O
YES
Aromatic or NH
YES
Alkene
YES
Aromatic
Examine region between 1500-1100 cm-1
11. Absorption ?
YES
12. Intense 1050-1300 cm-1?
NO
13. Medium Sharp 1375 cm-1?
NO
14. Medium Sharp 1450 cm-1?
NO
NO
No C-O, C-N, CH2, CH3
YES
C-C, C-O, C-N
YES
CH3
YES
CH2 or CH3
Examine region below 900 cm-1
15. Strong Absorptions?
YES
16. Sharp 720 cm-1?
NO
Aromatic, alkene or monochloro (C-Cl) possible
If Aromatic
Aromatic Substitution Patterns
Mono: 710-690 & 770-730 cm-1
1,2 Di: 770-735 cm-1
1,3 Di: 735-680 & 810-750 cm-1
1,4 Di: 860-800 cm-1
YES
-CH2- possible
NO
END

INFRARED SPECTRAL ANALYSIS WORKSHEET

Title: ____

Date:____________ Name (print)______________________________

ID#__________________________

Look at Region above 2700 cm^{-1}

Question	Answer	Conclusion (if any)
1. Absorption above 2700 cm^{-1}?		
2. Very broad and intense?		
3. Weak between 3100-3000 cm^{-1}?		
4. Sharp between 3000-2700 cm^{-1}?		
5. Medium sharp between 3300-3500 cm^{-1}?		

Look at Region between 2000-1500 cm^{-1} (if YES for Question 7: check for aldehyde C-H at 2700 cm^{-1})

Question	Answer	Conclusion (if any)
6. Absorption between 2000-1500 cm^{-1}?		
7. Intense 1660-1770 cm^{-1}?		
8. Sharp, medium & close to 1500-1600 cm^{-1}?		
9. Sharp, medium 1640-1840 cm^{-1}?		
10. Weak absorbances between 1770-2000 cm^{-1}?		

Look at Region between 1500-1100 cm^{-1} (peaks in this region are not easy to assign. Inferences should be confirmed from othersources!)

Question	Answer	Conclusion (if any)
11. Absorption between 1500-1100 cm^{-1}?		
12. Intense between 1300-1050 cm^{-1}?		
13. Medium sharp 1375 cm^{-1}?		
14. Medium sharp 1450 cm^{-1}?		

Look at Region below 900 cm^{-1} (peaks in this region are not easy to assign. Inferences should be confirmed from other sources!)

Question	Answer	Conclusion (if any)
15. Strong absorptions below 900 cm^{-1}?		
16. Sharp at 720 cm^{-1}?		

INFRARED SPECTRAL ANALYSIS WORKSHEET

Title: ____

Date:____________ Name (print)________________________________

ID#________________________

Look at Region above 2700 cm^{-1}

Question	Answer	Conclusion (if any)
1. Absorption above 2700 cm^{-1}?		
2. Very broad and intense?		
3. Weak between 3100-3000 cm^{-1}?		
4. Sharp between 3000-2700 cm^{-1}?		
5. Medium sharp between 3300-3500 cm^{-1}?		

Look at Region between 2000-1500 cm^{-1} (if YES for Question 7: check for aldehyde C-H at 2700 cm^{-1})

Question	Answer	Conclusion (if any)
6. Absorption between 2000-1500 cm^{-1}?		
7. Intense 1660-1770 cm^{-1}?		
8. Sharp, medium & close to 1500-1600 cm^{-1}?		
9. Sharp, medium 1640-1840 cm^{-1}?		
10. Weak absorbances between 1770-2000 cm^{-1}?		

Look at Region between 1500-1100 cm^{-1} (peaks in this region are not easy to assign. Inferences should be confirmed from other sources!)

Question	Answer	Conclusion (if any)
11. Absorption between 1500-1100 cm^{-1}?		
12. Intense between 1300-1050 cm^{-1}?		
13. Medium sharp 1375 cm^{-1}?		
14. Medium sharp 1450 cm^{-1}?		

Look at Region below 900 cm^{-1} (peaks in this region are not easy to assign. Inferences should be confirmed from other sources!)

Question	Answer	Conclusion (if any)
15. Strong absorptions below 900 cm^{-1}?		
16. Sharp at 720 cm^{-1}?		

Chemistry 145
Unknown Report 1

Name: Last, First ID #

Circle One: LIQUID SOLID MP/BP (°C) value or range:

Flame Tests: Conclusion

Combustion:	+(sooty)	-(clean)	
Beilstein:	+(green)	-	

Solubility Tests: Chemical Equation (if any)

H_2O:	soluble	insoluble	
NaOH:	soluble	insoluble	
$NaHCO_3$:	soluble	insoluble	
HCl:	soluble	insoluble	
conc H_2SO_4:	soluble	insoluble	

Color Reactions: Appropriate Chemical Reaction (indicate any visual observation)

2,4-DNP:	+	-	
I_2/NaOH:	+	-	
$Ag(NH_3)_2OH$:	+	-	
Br_2/H_2O:	+	-	
Br_2/CCl_4:	+	-	
H_2CrO_4/H_2SO_4:	+	-	
$KMnO_4$/NaOH	+	-	
$AgNO_3$/EtOH	+	-	

IR Absorption: Frequency (cm^{-1}) Structural Feature/Functional Group

General Formula for the unknown:

Chemistry 145

Unknown Report 1 — Name: Last, First — ID #

Circle One: LIQUID SOLID MP/BP (°C) value or range:

Flame Tests: Conclusion

Combustion:	+(sooty)	-(clean)	
Beilstein:	+(green)	-	

Solubility Tests: Chemical Equation (if any)

H_2O:	soluble	insoluble	
NaOH:	soluble	insoluble	
$NaHCO_3$:	soluble	insoluble	
HCl:	soluble	insoluble	
conc H_2SO_4:	soluble	insoluble	

Color Reactions: Appropriate Chemical Reaction (indicate any visual observation)

2,4-DNP:	+	-	
I_2/NaOH:	+	-	
$Ag(NH_3)_2OH$:	+	-	
Br_2/H_2O:	+	-	
Br_2/CCl_4:	+	-	
H_2CrO_4/H_2SO_4:	+	-	
$KMnO_4$/NaOH	+	-	
$AgNO_3$/EtOH	+	-	

IR Absorption: Frequency (cm^{-1}) Structural Feature/Functional Group

General Formula for the unknown:

Chemistry 145
Unknown Report 2

Name: Last, First ID #

Circle One: LIQUID SOLID

Unknown NAME:

Unknown STRUCTURE:

Unknown Molecular Formula:

^{1}H-NMR Data:

Chemical Shift (δ) (ppm)	Intensity Measured, Relative	Splitting	Structural Feature

Structure of base peak ion in mass spectrum:

MW (from MS):

MW Calculated (XXX.XX):

MP/BP Observed:

MP/BP Literature Value:

Chemistry 145
Unknown Report 2

Name: Last, First ID #

Circle One: LIQUID SOLID

Unknown NAME:

Unknown STRUCTURE:

Unknown Molecular Formula:

^{1}H-NMR Data:

Chemical Shift (δ) (ppm)	Intensity Measured, Relative	Splitting	Structural Feature

Structure of base peak ion in mass spectrum:

MW (from MS):

MW Calculated (XXX.XX):

MP/BP Observed:

MP/BP Literature Value:

MELTING POINTS & BOILING POINTS OF SELECTED ORGANIC COMPOUNDS

ALCOHOLS

bp (°C)	mp (°C)	Compound
65		Methanol
78		Ethanol
82		2-Propanol
83		i-Butyl alcohol
83		t-Butyl alcohol
96-98		Allyl alcohol
97		1-Propanol
98		2-Butanol
102		2-Methyl-2-butanol
104		2-Methyl-3-butyn-2-ol
108		2-Methyl-l-propanol
114		3-Methyl-2-butanol
114-115		Propargyl alcohol
114-115		3-Pentanol
118		1-Butanol
118-119		2-Pentanol
120		3,3-Dimethyl-2-butanol
120		2,3-Dimethyl-2-butanol
123		3-Methyl-3-pentanol
123		2-Methyl-2-pentanol
125		2-Methoxyethanol
128		2-Methyl-3-pentanol
129		2-Chloroethanol
130		2-Methyl-1-butanol
132		4-Methyl-2-pentanol
132		3-Methyl-1-butanol
135		2-Ethoxyethanol
136		3-Hexanol
136-138		1-Pentanol
139		2-Hexanol
139-140		Cyclopentanol
140		2,4-Dimethyl-3-pentanol

ALCOHOLS (con't)

bp (°C)	mp (°C)	Compound
146		2-Ethyl-1-butanol
151		2,2,2-Trichloroethanol
156		4-Heptanol
157		1-Hexanol
159		2-Heptanol
160-161		Cyclohexanol
161		3-Chloro-1-propanol
170		Furfuryl alcohol
176		1-Heptanol
178		2-Octanol
178		Tetrahydrofurfuryl alcohol
183-184		2,3-Butanediol
183-186		2-Ethyl-l-hexanol
187		1,2-Propanediol
194-197		Linalool
195		1-Octanol
196-198		Ethylene glycol
204		1,3-Butanediol
203-205		Benzyl alcohol
204		1-Phenylethanol
219-221		2-Phenylethanol
230		1,4-Butanediol
231		1-Decanol
237		2-Phenoxyethanol
237		3-Phenyl-1-propanol
259		4-Methoxybenzyl alcohol
258	33-35	Cinnamyl alcohol
	38-40	1-Tetradecanol
	44	Menthol
	48-50	1-Hexadecanol
	52	2,2-Dimethyl-1-propanol
	58-60	1-Octadecanol
217	60	4-Methylbenzyl alcohol
	66-67	Benzhydrol
	137	Benzoin (2-hydroxy-2-phenylacetophenone)

ALCOHOLS (con't)

bp (°C)	mp (°C)	Compound
	147	Cholesterol
	161	Triphenylmethanol

ALDEHYDES

bp (°C)	mp (°C)	Compound
21		Acetaldehyde
46-50		Propionaldehyde
63		Isobutyraldehyde
68		2-Methyl-2-propenal
75		Butyraldehyde
75		Trimethylacetaldehyde
90-92		3-Methylbutanal
98		Chloral
103		Pentanal
104		Crotonaldehyde
117		2-Ethylbutanal
121		4-Methylpentanal
131		Hexanal
144		5-Methylhexanal
153		Heptaldehyde
162		2-Furaldehyde
163		2-Ethylhexanal
170		Octanal
179		Benzaldehyde
195		Phenylacetaldehyde
197		Salicylaldehyde
199		3-Tolualdehyde
200		2-Tolualdehyde
204-205		4-Tolualdehyde
209-215		2-Chlorobenzaldehyde
209-215		3-Chlorobenzaldehyde

ALDEHYDES (con't)

bp (ºC)	mp (ºC)	Compound
230		3-Methoxybenzaldehyde
234		3-Bromobenzaldehyde
247		2-Ethoxybenzaldehyde
248		4-Anisaldehyde
250-252		E-Cinnamaldehyde
	33-34	1-Naphthaldehyde
238	37-39	2-Anisaldehyde
	42-45	3,4-Dimethoxybenzaidehyde
	44	3,4-Dichlorobenzaldehyde
	44-47	4-Chlorobenzaldehyde
	49-52	2,5-Dimethoxybenzaldehyde
	54	2,3-Dimethoxybenzaldehyde
	57-59	3-Nitrobenzaldehyde
	65	3,5-Dichlorobenzaldehyde
	71	2,6-Dichlorobenzaldehyde
	71	2,4-Dimethoxybenzaldehyde
	72-75	p-(Dimethylamino)benzaldehyde
	78	3,4,5-Trimethoxybenzaldehyde
	81-83	Vanillin (4-hydroxy-3-methoxybenzaldehyde)
	104	3-Hydroxybenzaldehyde
	116	4-Hydroxybenzaldehyde
	118	2,4,6-Trimethoxybenzaldehyde

AMIDES

bp (ºC)	mp (ºC)	Compound
153		N,N-Dimethylformamide
164-166		N,N-Dimethylacetamide
178		N,N-Diethylformamide
185		N-Methylformamide
186		N,N-Diethylacetamide
199		N-Ethylformamide

AMIDES (con't)

bp (°C)	mp (°C)	Compound
205		N-Ethylacetamide
210		Formamide
243-244		N-Methylformanilide
	26-28	N-Methylacetamide
	54	N-Ethylacetanalide
	60	N-Benzylformamide
	79-81	Acetamide
	109-111	Methacrylamide
	113-115	Acetanilide
	116-118	2-Chloroacetamide
	162-164	Benzanilide
	177-179	4-Chloroacetanilide

AMINES (PRIMARY & SECONDARY)

bp (°C)	mp (°C)	Compound
33-34		Isopropylamine
36		Ethylmethylamine
46		t-Butylamine
48		n-Propylamine
53		Allylamine
55		Diethylamine
63		s-Butylamine
64-71		Isobutylamine
78		n-Butylamine
84		Di-isopropylamine
87-88		Pyrrolidine
89		Triethylamine
96		Isopentylamine
104		n-Pentylamine
106		Piperidine

AMINES (PRIMARY & SECONDARY, con't)

bp (°C)	mp (°C)	Compound
111		Di-n-propylamine
118		Ethylenediamine
129		Morpholine
132		n-Hexylamine
134		Cyclohexylamine
137-139		Di-isobutylamine
145-146		Furfurylamine
149		N-Methylcyclohexylamine
156		n-Heptylamine
159		Di-n-butylamine
182-185		Benzylamine
184		Aniline
187		1-Amino-1-phenylethane
196		N-Methylaniline
198		2-Amino-1-phenylethane
199-200		2-Toluidine
203-204		3-Toluidine
205		N-Ethylaniline
208-210		2-Chloroaniline
210		2-Ethylaniline
216		2,6-Dimethylaniline
216		4-Ethylaniline
218		2,4-Dimethylaniline
218		2,5-Dimethylaniline
221		N-Ethyl-m-toluidine
225		2-Anisidine
230		3-Chloroaniline
231-233		2-Phenetidine
241		4-Chloro-2-methylaniline
242		3-Chloro-4-methylaniline
250		4-Phenetidine
256		Dicyclohexylamine
	35-38	N-Phenylbenzylamine
	41-44	4-Toluidine
	49-51	2,5-Dichloroaniline

AMINES (PRIMARY & SECONDARY, con't)

bp (°C)	mp (°C)	Compound
	52-54	Diphenylamine
	57-60	4-Anisidine
	57-60	2-Aminopyridine
	60-62	N-Phenyl-1-naphthylamine
	62-65	2,4,5-Trimethylaniline
	64-66	1,3-Phenylenediamine
	66	4-Bromoaniline
	68-71	4-Chloroaniline
	71-73	2-Nitroaniline
	97-99	2,4-Diaminotoluene
	100-102	1,2-Phenylenediamine
	104-107	2-Methyl-5-nitroaniline
	107-109	2-Chloro-4-nitroaniline
	112-114	3-Nitroaniline
	115-116	4-Methyl-2-nitroaniline
	117-119	4-Chloro-2-nitroaniline
	120-122	2,4,6-Tribromoaniline
	131-133	2-Methyl-4-nitroaniline
	138-140	2-Methoxy-4-nitroaniline
	138-142	1,4-Phenylenediamine
	148-149	4-Nitroaniline
	162-164	4-Aminoacetanilide
	176-178	2,4-Dinitroaniline

AMINES (TERTIARY)

bp (°C)	mp (°C)	Compound
85-91		Triethylamine
115		Pyridine
128-129		2-Picoline

AMINES (TERTIARY, con't)

bp (°C)	mp (°C)	Compound
143-145		2,6-Lutidine
144		3-Picoline
145		4-Picoline
155-158		Tri-n-propylamine
159		2,4-Lutidine
183-184		N,N-Dimethylbenzylamine
193		N,N-Dimethylanaline
216		Tri-n-butylamine
217		N,N-Diethylaniline
237		Quinoline

CARBOXYLIC ACIDS

bp (°C)	mp (°C)	Compound
101		Formic acid
118		Acetic acid
139		Acrylic acid
141		Propionic acid
154		Isobutyric (2-methylpropionic) acid
162		n-Butyric acid
163		Methacrylic (or Vinylacetic) acid
163	34	Pivalic (Trimethylacetic) acid
164		3-Butenoic (vinylacetic) acid
165		Pyruvic acid
169		Isocrotonic acid
174		2-Methylbutanoic acid
176		Isovaleric (3-methylbutanoic) acid
181		Crotonic acid
185		Valeric acid
185		tert-Butylacetic acid

CARBOXYLIC ACIDS (con't)

bp (°C)	mp (°C)	Compound
186		2-Chloropropinic acid
186		2-Methylvaleric acid
186		2,2-Dimethylbutanoic acid
186		2,3-Dimethylbutanoic acid
194		Dichloroacetic acid
195		2-Ethylbutanoic (diethylacetic) acid
202-203		Hexanoic acid
206		Ethoxyacetic acid
223		Heptanoic acid
228		2-Ethylhexanoic acid
237		Octanoic acid
254		Nonanoic acid
	31-32	Decanoic acid
	37	Acetoacetic acid
	42	2-Phenylbutanoic acid
	43-45	Lauric acid
	47-49	Bromoacetic acid
	47-49	Hydrocinnamic acid
	54-55	Myristic acid
	54-58	Trichloroacetic acid
	61-62	Chloroacetic acid
	61-62.5	Palmitic acid
	67-69	Stearic acid
	68-69	3,3-Dimethylacrylic acid
	71-73	Crotonic acid
	77-78.5	Phenylacetic acid
	87	4-Methoxyphenylacetic acid
	98	3-Phenoxypropionic acid
	99	Phenoxyacetic acid
	101-102	Oxalic acid dihydrate
	98-100	2-Anisic acid
	98-102	Azelaic acid (nonanedioic)
	103-105	o-Toluic acid
	104	4-Chlorophenylacetic acid
	108-110	m-Toluic acid

CARBOXYLIC ACIDS (con't)

bp (°C)	mp (°C)	Compound
	110	3-Methoxybenzoic (m-anisic) acid
	119-121	DL-Mandelic acid
	122-123	Benzoic acid
	127-128	2-Benzoylbenzoic acid
	129-130	2-Furoic acid
	131-133	DL-Malic acid
	131-134	Sebacic acid
	134-135	E-Cinnamic acid
	134-136	Maleic acid
	135-137	Malonic acid
	138-140	Acetylsalicylic acid
	138-140	2-Chlorobenzoic acid
	140-142	3-Nitrobenzoic acid
	141	2,4-Dichlorophenoxyacetic acid
	144	2,4,5-Trimethoxybenzoic acid
	144-148	Anthranilic acid
	146	2-Chlorophenoxyacetic acid
	147-149	Diphenylacetic acid
	152-153	Adipic acid
	153-154	Citric acid
	155-157	3-Chlorobenzoic acid
	157-159	4-Chlorophenoxyacetic acid
	157-160	2,4-Dichlorobenzoic acid
	158-160	Salicylic acid
	163-164	Trimethylacetic acid
	164-166	5-Bromosalicylic acid
	166-167	Itaconic acid
	168-171	3,4,5-Trimethoxybenzoic acid
	171-174	D-Tartaric acid
	179-182	3,4-Dimethoxybenzoic acid
	180-182	4-Toluic acid
	182-185	4-Anisic acid
	187-190	Succinic acid
	188-189	4-Aminobenzoic acid
	201-203	3-Hydroxybenzoic acid

CARBOXYLIC ACIDS (con't)

bp (°C)	mp (°C)	Compound
	203-206	3,5-Dinitrobenzoic acid
	207-209	3,4-Dichlorobenzoic acid
	210-211	Phthalic acid
	214-215	4-Hydroxybenzoic acid
	225-227	2,4-Dihydroxybenzoic acid
	236-239	Nicotinic acid
	239-241	4-Nitrobenzoic acid
	239-241	4-Chlorobenzoic acid
	299-300	Fumaric acid

ESTERS

bp (°C)	mp (°C)	Compound
34		Methyl formate
52-54		Ethyl formate
57		Methyl acetate
71		Isopropyl formate
72-73		Vinyl acetate
77		Ethyl acetate
79		Methyl propionate
80		Methyl acrylate
81		Propyl formate
85		Isopropyl acetate
93		Ethyl chloroformate
93		Methyl isobutyrate
94		Isopropenyl acetate
98		Isobutyl formate
98		t-Butyl acetate
99		Ethyl propionate
99		Ethyl acrylate
100		Methyl methacrylate
101		Methyl trimethylacetate

ESTERS (con't)

bp (°C)	mp (°C)	Compound
102		Methyl butyrate
102		n-Propyl acetate
106-113		s-Butyl acetate
110		Ethyl isobutyrate
110		Isopropyl propionate
117		Isobutyl acetate
117		Methyl isovalerate
118		Ethyl trimethylacetate
120		Ethyl butyrate
123		Propyl propionate
127		n-Butyl acetate
128		Methyl valerate
130		Methyl chloroacetate
130		Methyl methoxyacetate
131-133		Ethyl isovalerate
137		Methyl pyruvate
142		n-Amyl acetate
142		Isoamyl acetate
143		Ethyl chloroacetate
154		Ethyl lactate
155		Ethyl pyruvate
158		Ethyl dichloroacetate
167		Butyl butyrate
168		Ethyl caproate (ethyl hexanoate)
168		Ethyl trichloroacetate
170		Methyl acetoacetate
172		Hexyl acetate
175		Cyclohexyl acetate
181		Ethyl acetoacetate
182		Dimethyl malonate
185		Diethyl oxylate
196		Dimethyl succinate
196		Dimethyl methylsuccinate
197		Phenyl acetate
199		Diethyl malonate

ESTERS (con't)

bp (°C)	mp (°C)	Compound
199		Methyl benzoate
213		Ethyl benzoate
217		Benzyl acetate
218		Diethyl fumarate
218		Diethyl succinate
218		Isopropyl benzoate
220		Methyl phenylacetate
224		Methyl salicylate
228		Ethyl phenylacetate
230		Propyl benzoate
234		Ethyl salicylate
	30	Ethyl 2-nitrobenzoate
	33	Methyl p-toluate
	36	Methyl cinnamate
	39	Benzyl cinnamate
	42	Phenyl salicylate
	47	Ethyl 3-nitrobenzoate
	52	Methyl 2-hydroxy-2-phenylacetate
	56	Ethyl 4-nitrobenzoate
	69	Phenyl benzoate
	70	Methyl 3-hydroxybenzoate
	78	Methyl 3-nitrobenzoate
	90	Dimethyl tartarate
	90	Ethyl 4-aminobenzoate
	96	Methyl 4-nitrobenzoate
	96	Propyl 4-hydroxybenzoate
	102	Dimethyl fumarate
	116	Ethyl 4-hydroxybenzoate
	131	Methyl 4-hydroxybenzoate
	136	Acetylsalicylic acid

KETONES

bp (ºC)	mp (ºC)	Compound
56		Acetone
80		2-Butanone
88		2,3-Butanedione
94		3-Methyl-2-butanone
100-101		2-Pentanone
102		3-Pentanone
106		Pinacolone (3,3-dimethyl-2-butanone)
114-116		4-Methyl-2-pentanone
115		1-Methyl-2-propanone
117		4-Methyl-2-pentanone
118		3-Methyl-2-pentanone
124		2,4-Dimethyl-3-pentanone
125		4,4-Dimethyl-2-pentanone
125		3-Hexanone
128		2-Hexanone
128-129		5-Hexen-2-one
129		4-Methyl-3-penten-2-one
130-131		Cyclopentanone
133-135		2,3-Pentanedione
136		2-Methyl-3-hexanone
136		5-Methyl-3-hexanone
137		Methyl pyruvate
145		4-Heptanone
145		5-Methyl-2-hexanone
145-147		2-Heptanone
146-149		3-Heptanone
155		Ethyl pyruvate
156		Cyclohexanone
162		3,5-Dimethyl-4-heptanone
162-163		2-Methylcyclohexanone
169		2,6-Dimethyl-4-heptanone
169-170		3-Methylcyclohexanone
170		4-Octanone
173		2-Octanone

Ketones (con't)

bp (°C)	mp (°C)	Compound
181		Ethyl acetoacetate
191		Acetonylacetone (2,6-hexandione)
202		Acetophenone
216		Phenylacetone
217		Isobutyrophenone
218		Propiophenone
226		4-Methylacetophenone
228-230		n-Butyrophenone
231-232		2-Undecanone
232		4-Chloroacetophenone
235		Benzylacetone
	35-37	4-Chloropropiophenone
	35-39	4-Phenyl-3-buten-2-one
	36-38	4-Methoxyacetophenone
	39	2-Methoxybenzophenone
	48-49	Benzophenone
	51	3,4-Dimethoxyacetophenone
	53-55	2-Acetonaphthone
	57	4-Methylbenzophenone
	60	Desoxybenzoin
	62	4-Methoxybenzophenone
	76-78	3-Nitroacetophenone
	78	4-Chlorobenzophenone
	78-80	4-Nitroacetophenone
	82-85	9-Fluorenone
	137	Benzoin (2-hydroxy-2-phenylacetophenone)
	147-148	4-Hydroxypropiophenone

NITRILES

bp (°C)	mp (°C)	Compound
77		Acrylonitrile
81		Acetonitrile
83-84		Trichloroacetonitrile
97		Propionitrile
107-108		Isobutyronitrile
115-117		n-Butyronitrile
174-176		3-Chloropropionitrile
191		Benzonitrile
205		2-Tolunitrile
213		3-Tolunitrile
233-234		Benzyl cyanide

PHENOLS

bp (°C)	mp (°C)	Compound
175-176		2-Chlorophenol
181	42	Phenol
191	32-33	o-Cresol
197		Salicylaldehyde
202	32-34	p-Cresol
203		m-Cresol
205	32	2-Methoxyphenol
207		2-Ethylphenol
212	27	2,4-Dimethylphenol
224		Methyl salicylate
228-229		3,4-Dimethylphenol
236	32	3-Bromophenol
243		3-Methoxyphenol
	36	2,4-Dibromophenol
	36	4-Methyl-2-nitrophenol

PHENOLS (con't)

bp (°C)	mp (°C)	Compound
	42	Phenol
	42	Phenyl salicylate
	42-43	2,4-Dichlorophenol
218-219	42-45	4-Ethylphenol
	43-45	4-Chlorophenol
	44-46	2,6-Dimethylphenol
	44-46	2-Nitrophenol
	49	4-Chloro-2-methylphenol
	49-51	Thymol (5-methyl-2-isopropylphenol)
	53-56	5-Methyl-2-nitrophenol
	55	4-Methoxyphenol
	59	2,5-Dichlorophenol
	60	4-Isopropylphenol
	62-64	3,5-Dimethylphenol
	63	3,4-Dimethylphenol
	64-68	4-Bromophenol
	66	4-Chloro-3-methylphenol
	67	2,6-Dichlorophenol
	68	3,4-Dichlorophenol
	68	3,5-Dichlorophenol
	68	2,4,5-Trichlorophenol
	70	Methyl 3-hydroxybenzoate
	71-74	2,4,6-Trimethylphenol
	74	2,5-Dimethylphenol
	81	Vanillin (4-hydroxy-3-methoxybenzaldehyde)
	92-95	2,3,5-Trimethylphenol
	95-96	1-Naphthol
	96	Propyl 4-hydroxybenzoate
	97	3-Nitrophenol
	98-101	4-t-Butylphenol
	104-105	Catechol
	109-110	Resorcinol
	112-114	4-Nitrophenol
	114	2,4-Dinitrophenol
	116-118	Ethyl 4-hydroxybenzoate

PHENOLS (con't)

bp (°C)	mp (°C)	Compound
	117	4-Hydroxybenzaldehyde
	121-124	2-Naphthol
	126-128	Methyl 4-hydroxybenzoate
	129	3-Methyl-4-nitrophenol
	133-134	Pyrogallol
	158	Salicylic acid
	173	3,5-Dinitrosalicylic acid
	200	3-Hydroxybenzoic acid
	215	4-Hydroxybenzoic acid

Laboratory Check-Out, Chem. 145

A. Remove all equipment from your drawer. Make sure each item is clean and dry (wash and remove labels if necessary).
B. Obtain a replacement for any broken or cracked item.
C. Have a TA or the Instructor verify each item before you put it back in your drawer. Have a TA or the Instructor sign this checklist and then lock your drawer with a Chemistry Department lock.
D. Give the signed Checklist to the Instructor **to insure you receive all points** for checking out.
E. Turn in your laboratory notebook to the Instructor if you are told to do so.
F. Fill out a Course Evaluation Form and return the completed form to the envelope provided.

Equipment
(**Bold** numbers refer to illustration of glassware)

1 beakers, 100 mL (2)
1 beakers, 250 mL (2)
1 beaker, 1000 mL (1)
2 Erlenmeyer flasks, 125 mL (2)
2 Erlenmeyer flasks, 250 mL (2)
3 filter flask, 250 mL (1)
4 graduated cylinder, 25 mL (1)
5 graduated cylinder, 100 mL (1)
6 glass funnels (Bunsen (1) and fluted (1), one may have short stem)
7 Buchner funnel (porcelain, with rubber stopper) (1)
8 separatory funnel, 250 ml (including stopper)

20 wire gauze (1)
21 glass stirring rod (1)
spatula (1)
thermometer, 250°C (1)
test tube rack (1)
test tubes, 3 in. (2)
test tubes, 6 in. (6)
test tubes, 8 in. (2)
goggles (1 pr)

cloth towel (1)
misc. corks, pipettes, vials

19/22 standard taper glassware (1 ea)
9 round bottom flask, 25 mL
9 round bottom flask, 50 mL
9 round bottom flask, 100 mL
10 vacuum adapter
11 connecting adapter
12 outlet adapter
12 neoprene fitment
13 Claisen adapter
14 West condenser (thin)
15 distillation column (fat)
16 separatory funnel, 125 mL
17 stopper, 19/22
18 round bottom flask, 250 mL, with side
19 Keck clamps (blue plastic) (2)

________________________ Student Name (Print)

________________________ TA or Instructor (signature)

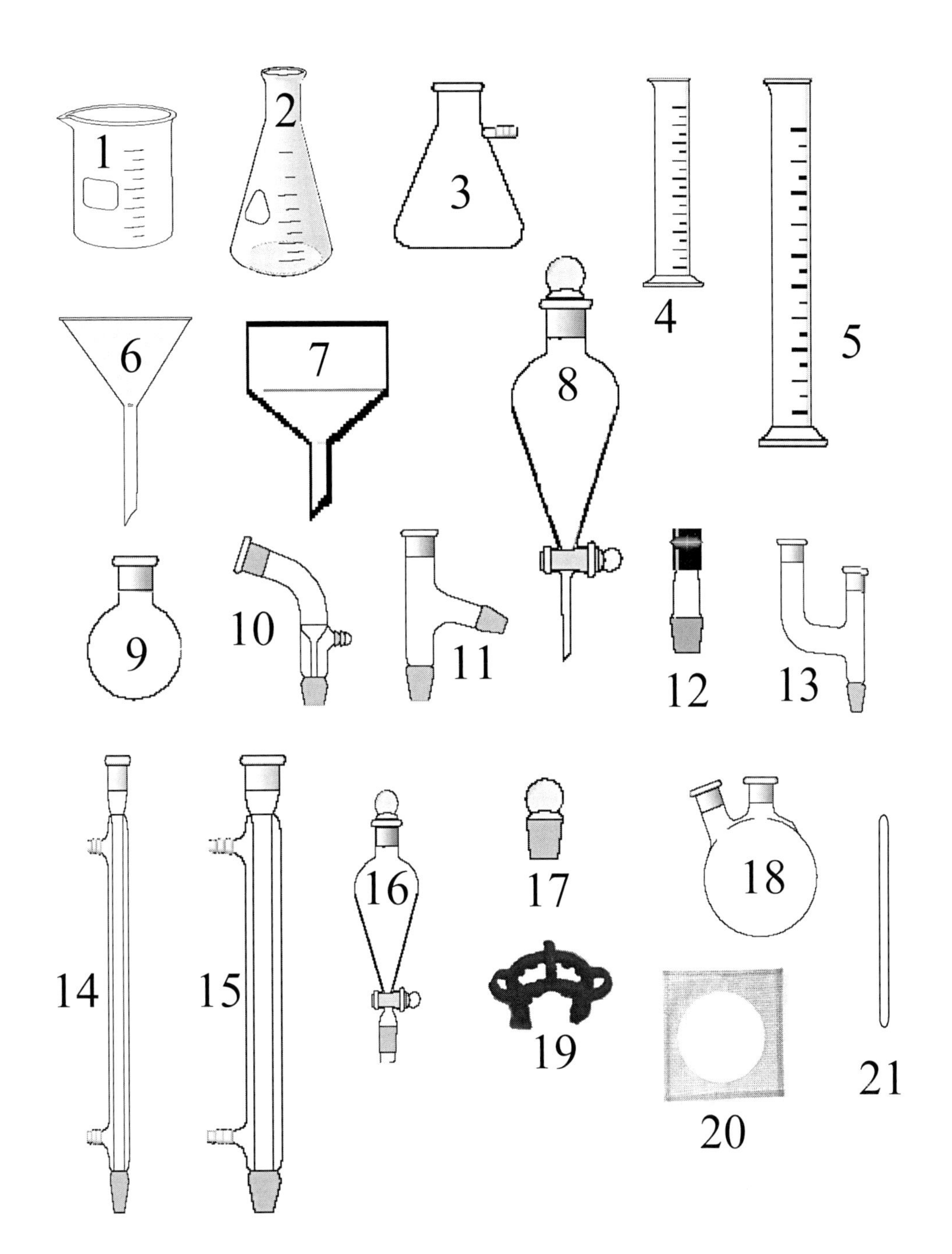
1
2
3
4
5
6
7
8
9
10
11
12
13
14
15
16
17
18
19
20
21

PROPERTIES OF COMMON SOLVENTS

Name	MP (°C)	BP (°C)	Refractive Index	Density (g/ml)	Solubility in H_2O, g/100 mL	Dipole Moment μ	Dielectric Constant ϵ
Hydrocarbons							
Pentane	-130	36	1.3580	0.626	0.036	0.0	1.84
Hexane	- 95	69	1.3748	0.659	insol	0.0	1.89
Cyclohexane	6.5	81	1.4255	0.779	insol	0.0	2.02
Heptane	- 91	98	1.3870	0.684	insol	0.0	1.98
Methyl-cyclohexane	-126	101	1.4222	0.77	insol	0.0	2.02
Benzene	5.5	80	1.5007	0.879	0.5	0.0	2.28
Toluene	- 93	111	1.4963	0.865	v.sl.	0.4	2.38
o-Xylene	- 24	144	1.5048	0.897	insol	0.6	2.57
Ethylbenzene	- 95	136	1.4952	0.867	insol	0.6	2.41
p-Xylene	12	138	1.4954	0.866	insol	0	2.27
Ethers							
Diethyl ether	-116	35	1.3506	0.715	7.0	1.2	4.34
Tetrahydrofuran	-108	66	1.4070	0.887	misc	1.6	7.32
Dimethoxy-ethane	- 69	85	1.3790	0.867	misc	-	-
Dioxane	12	101	1.4206	1.034	misc	0.0	2.21
Dibutyl ether	- 98	142	1.3988	0.764	insol	-	-
Anisole	- 37	154	1.5160	0.995	insol	1.4	4.33
Diglyme	- 64	160	1.4073	0.937	misc	-	-
Chloro-hydrocarbons							
Dichloromethane	- 97	40	1.4240	1.325	2.0	1.6	8.9
Chloroform	- 63	61	1.4453	1.492	0.5	1.9	4.7

Name	MP (°C)	BP (°C)	Refractive Index	Density (g/ml)	Solubility in H_2O, g/100 mL	Dipole Moment μ	Dielectric Constant ϵ
Carbon tetrachloride	- 23	77	1.4595	1.594	0.025	0.0	2.23
1,2-Dichloro-ethane	- 35	83	1.4438	1.256	0.9	2.1	10.0
Chlorobenzene	- 46	132	1.5236	1.106	insol	1.7	5.62
1,2-Dichloro-benzene	- 17	178	1.5504	1.305	insol	2.5	9.93
Alcohols							
Methanol	- 98	65	1.3280	0.791	misc	1.7	32.6
Ethanol	-114	78.5	1.361	0.798	misc	1.7	24.3
2-Propanol	- 90	82	1.3770	0.785	misc	1.7	18.3
tert-Butanol	25	83	1.3860	0.786	misc	1.7	10.9
n-Propanol	-127	97	1.3840	0.804	misc	1.7	20.1
n-Butanol	- 90	118	1.3985	0.810	9.1	1.7	17.1
2-Methoxy-ethanol	- 85	124	1.4020	0.965	misc	2.2	16.0
2-Ethoxyethanol	- 90	135	1.4068	0.930	misc	2.1	-
Ethylene glycol	- 13	198	1.4310	1.113	misc	2.3	37.7
Dipolar Aprotic							
Acetone	- 94	56	1.3585	0.791	misc	2.9	20.7
Acetonitrile	- 44	81	1.3440	0.786	misc	3.94	36.2
Nitromethane	- 29	101	1.3820	1.137	9.1	3.46	38.6
Dimethyl-formamide	- 61	153	1.4305	0.944	misc	3.7	36.7
Dimethyl sulfoxide	18	189	1.4780	1.101	misc	3.96	47
N,N-Dimethyl-acetamide	- 20	165	1.4375	0.937	misc	3.8	37.8
Formamide	2	210	1.4440	1.134	misc	3.7	110

Name	MP (°C)	BP (°C)	Refractive Index	Density (g/ml)	Solubility in H_2O, g/100 mL	Dipole Moment μ	Dielectric Constant ϵ
Hexamethylphos-phoramide	7	230	1.4579	1.030	misc	-	-
Tetramethylene sulfone	27	285	1.4840	1.261	misc	4.7	44
Miscellaneous							
Carbon Disulfide	-112	46	1.6270	1.266	0.3	0.0	2.64
Ethyl acetate	-84	76	1.3720	0.902	10.	1.8	6.
Methyl ethyl ketone	-86	80	1.3780	0.805	27.5	2.5	18.5
Water	0	100	1.330	1.000	-	1.8	78.4
Formic Acid	8.5	101	1.3721	1.220	misc	1.41	58
Pyridine	- 42	115	1.5090	0.978	misc	2.19	12.3
Nitrobenzene	5	210	1.5513	1.204	0.2	4.01	35
Acetic acid	16	117	1.3720	1.049	misc	1.7	6.2

Abbreviations: insol, insoluble; v.sl., very slightly; misc, miscible in all proportions; aq, aqueous.

CONCENTRATED ACIDS AND BASES

Name	Chemical Formula	MW	% by Weight	Specific Gravity	Molarity
Acetic Acid (glacial)	CH_3CO_2H	60.05	99.8	1.05	17.4
Ammonium Hydroxide (as Ammonia)	NH_3	17.03	28	0.90	14.8
Hydrochloric Acid	HCl	36.46	37	1.19	12.0
Nitric Acid	HNO_3	63.01	70	1.42	15.9
Phosphoric Acid	H_3PO_4	97.99	85	1.70	14.7
Sulfuric Acid	H_2SO_4	98.07	96	1.84	18.0

Calculations involving concentrated acids and bases:

$$Concentration\left(\tfrac{Moles}{L}\right) = \frac{\left((10)\cdot(\%\,byWeight)\cdot(SpecificGravity)\right)}{(MolecularWeight)}$$

Number of Moles in a volume V mL:

$$\#\ Moles = \frac{(V)\cdot\left(\dfrac{\%\ by\,Weight}{100}\right)\cdot(SpecificGravity)}{(MW)}$$

GAS CHROMATOGRAPHY CONDITIONS

	Distillation	Extraction	Cyclohexene	Cyclohexanone	Methyl Benzoate	Nitration	Grignard	Aldol
Coluumn[a]	5 mm wide bore fused silica capillary							
Stationary Phase[a]	Methyl Siloxane (HP-1)							
Carrier Gas[a]	He							
Head Pressure[a]	2 psi							
Detector[a]	FID							
Detector Temp. (°C)[a]	250							
Injector Temp. (°C)[a]	220							
Initial Oven Temp. (°C)	50	100	80	40	70	110	80	140
Initial Time (min)	2	1.2	2	6	1	1	0	0.5
Rate (°C/min)	0	40	0	0	15	50	40	50
Intermediate Oven Temp. (°C)	-	-	-	-	-	140	-	-
Time at Intermediate Oven Temp. (min)	-	-	-	-	-	1	-	-
Final Oven Temp. (°C)	50	180	80	40	150	200	245	300
Final Time (min)	0	0	0	0	0.4	1	0	1.0
Sample Solvent	none	CH_2Cl_2	none	none	none	CH_2Cl_2	CH_2Cl_2	CH_2Cl_2

[a]Same For All Experiments
[b]FID = Flame Ionization Detector